Lab Manual for

BiologyLabs On-Line

by

Robert Desharnais
Jeffrey Bell

Michael A. Palladino
Monmouth College

An imprint of Addison Wesley Longman, Inc.

San Francisco • Reading, Massachusetts • New York • Don Mills, Ontario
Harlow, England • Amsterdam • Madrid • Sydney • Mexico City

Executive Editor: Erin Mulligan
Project Manager: Anne Scanlan-Rohrer
Associate Producer: Maureen Kennedy
Prepress Supervisor: Vivian McDougal
Marketing Manager: Gay Meixel
Copy Editor: Jan McDearmon
Manufacturing Coordinator: Holly Bisso
Compositor: Michael Palladino
Printer: Courier Company
Cover Design: Vivian McDougal and Rachel Smith

References are made in this lab manual to Campbell, N.A., J.B. Reece, and L.G. Mitchell. *Biology*, Fifth Edition. Menlo Park, CA: Benjamin-Cummings, 1999.

Combined Labs ISBN 0-8053-7443-4

1 2 3 4 5 6 7 8 9 10—CRS—03 02 01 00 99

Addison Wesley Longman, Inc.
1301 Sansome Street
San Francisco, CA 94111

Contents

MitochondriaLab

<u>Background</u>

Enzymes are an important class of proteins that function as catalysts in living cells. You may have already used EnzymeLab to learn about basic principles of enzyme activity and the role of enzymes in **metabolism**. Cell metabolism is dependent on the combined actions of many different enzymes that are essential for the anabolic and catabolic reactions that must occur to maintain the physiology of a cell. It is important to understand that many reactions that require enzymes rarely involve just a single enzyme that works by an all-or-none process. This would be like trying to manufacture a car from all of its components in one sweeping motion! Instead, many biochemical reactions involve cascades of enzymatic reactions called metabolic pathways. In a metabolic pathway, several enzymes work in a sequential fashion to convert reactants into a product or products. At each step in a metabolic pathway, intermediate molecules (metabolites) are produced that serve as the **substrates** for subsequent enzymes in the pathway. One benefit of a metabolic pathway compared with a single-enzyme reaction is that a cell can often precisely regulate the amount of product generated by independently controlling the catalytic activity of certain enzymes in the pathway.

Intermediate molecules are often an important part of the control of a metabolic pathway. One way in which metabolic pathways can be regulated by intermediates involves a process called **allosteric regulation**. In this form of control, a molecule binds to a portion of an enzyme, other than the active site, and alters the shape of the active site to either stimulate or inhibit the enzyme. **Feedback inhibition** is another mechanism that cells can use to regulate a metabolic pathway. In feedback inhibition, a final or end-product of a reaction can inhibit enzymes in the metabolic pathway. This process allows a cell to carefully control the amount of end-product that it produces as a way to prevent excess accumulation and waste of an end-product. Frequently, the end-product inhibits or blocks the activity of an enzyme at one of the initial, rate-limiting steps in the pathway to prevent the unnecessary production of intermediates. This would be similar to a car manufacturer, whose car supply has exceeded the public's demand, stopping the auto assembly line at the first step rather than halfway through the assembly process to avoid producing half-completed cars.

One of the most essential metabolic pathways that occur in all living cells involves the enzymatic conversion of food molecules to produce energy in the form of a molecule called **adenosine triphosphate (ATP)**. ATP is a critical energy source universally used by all living cells. ATP powers many chemical reactions by providing the phosphate groups necessary for phosphorylation reactions. Phosphorylation reactions are catalyzed by a broad class of enzymes called kinases. Kinases can transfer a phosphate group from ATP onto other molecules. Phosphorylation of molecules such as proteins often results in a change in the shape of the protein, which activates the protein to perform a desired function. For example, muscle contraction requires that the contractile proteins in a muscle cell—actin and myosin—be phosphorylated, thus enabling them to slide along one another to produce shortening of the muscle cell.

Although lipids and proteins can undergo catabolism in cells to serve as metabolic fuels to power the reactions necessary for ATP synthesis, carbohydrates such as glucose are excellent sources of energy for producing ATP. Many plant and animal cells, including human cells, can convert glucose into ATP in the presence of oxygen by a set of reactions called **aerobic cellular respiration**. Other cells—for example, certain yeast cells, bacteria, and human skeletal muscle cells—can produce ATP from glucose in the absence of oxygen (**anaerobic** conditions) via reactions called **fermentation**. Cellular respiration produces a much higher total yield of ATP from one molecule of glucose than does fermentation, while producing a minimal number of waste products. In addition to ATP, other end-products produced include water and carbon dioxide, the primary waste product produced by cellular respiration. The summary equation showing the net yield of products created by the catabolism of one molecule of glucose ($C_6H_{12}O_6$) is shown below:

$$C_6H_{12}O_6 + 6\ O_2 \rightarrow 6\ H_2O + 6\ CO_2 + 36\ ATP$$

Glucose catabolism via cellular respiration can be grouped into three major metabolic stages: (1) **glycolysis;** (2) the **Krebs cycle**, also known as the citric acid cycle or tricarboxylic acid cycle (TCA cycle); and (3) the **electron transport chain** and **oxidative phosphorylation**. In eukaryotic cells, glycolysis occurs in the cytoplasm of the cell. The Krebs cycle occurs in the **mitochondrial matrix**, while the reactions of the electron transport chain and oxidative phosphorylation occur on the **cristae** of the mitochondrion. These pathways rely on oxidation-reduction reactions in which electrons are enzymatically removed (**oxidation**) from glucose and transferred (**reduction**) to electron acceptor molecules such as **nicotinamide adenine dinucleotide (NAD^+)**. Upon receiving electrons, NAD^+ is reduced to NADH, which functions as an electron carrier that supplies electrons to an electron transport chain in mitochondria that will ultimately power ATP synthesis in the reactions known as oxidative phosphorylation. The reactions of cellular respiration are summarized in the next three paragraphs.

Glycolysis involves ten enzymatic steps that ultimately result in the degradation of one molecule of glucose into two molecules of a three-carbon acid called pyruvate (pyruvic acid). Glycolysis also results in the net production of two ATP molecules and two molecules of NADH. These reactions can occur under aerobic or anaerobic conditions, but the yield of pyruvate and NADH is always the same. Before entering the Krebs cycle, each pyruvate is subsequently converted into a molecule called acetyl coenzyme A (**acetyl CoA**).

The Krebs cycle involves eight enzymatic steps that serve to oxidize both molecules of acetyl CoA. The primary result is the production of six molecules of NADH and two molecules of an electron carrier molecule similar to NADH called flavin adenine dinucleotide ($FADH_2$). These reactions were originally known as the tricarboxylic acid cycle or citric acid cycle because each enzymatic step results in the production of eight organic acids called carboxylic acids (citrate, isocitrate, α-ketoglutarate, succinyl CoA, succinate, fumarate, malate, and oxaloacetate) as intermediate molecules. Each of these molecules serves as a substrate for the next enzyme in the Krebs cycle, and some of these intermediates serve to regulate enzymes of cellular respiration. You will design experiments with some of these intermediates using MitochondriaLab to test the effects of each intermediate on the reactions of cellular respiration.

During the reactions of the electron transport chain, the electrons stored in the NADH and $FADH_2$ produced by the Krebs cycle are enzymatically transferred to a series of electron acceptor molecules in the cristae. Many of the electron acceptor molecules are iron-containing proteins called **cytochromes**. The cytochromes are clustered together within the cristae to form four multiprotein groups, or complexes of proteins—named groups I, II, III, and IV—that transfer electrons in a sequential fashion. Molecular oxygen serves as the final electron acceptor in this chain. Oxygen molecules receive electrons from complex IV. This reaction represents the oxygen-consuming stage of cellular respiration because reduced oxygen molecules are converted into water during this transfer. As electrons are transferred along this chain, this series of oxidation-reduction reactions generates a hydrogen ion (H^+) gradient (called a **proton-motive force**) in the intermembrane space of the mitochondrion. This H^+ gradient provides the energy necessary for an enzyme called **ATP synthase**. This enzyme also functions as an ion channel to allow H^+ flow down a gradient from the intermembrane space into the mitochondrial matrix. The H^+ flow through ATP synthase activates the enzyme to synthesize ATP from ADP and inorganic phosphate in a final stage called oxidative phosphorylation.

Studying the reactions of cellular respiration is an excellent way to develop an understanding of the complexities of cell metabolism and the mechanisms by which a cell can control its metabolism. One way to study the reactions of cellular respiration involves performing experiments on preparations of mitochondria that have been isolated from mitochondria-rich tissue, such as skeletal muscle or liver, using cell fractionation techniques. Many of the details of these reactions have been well understood for almost a hundred years. Numerous scientists were involved in carrying out the biochemical experiments that were used to learn about the substrates, enzymes, and cofactors required for cellular respiration, the sequence of each reaction relative to one another, and the regulation of these reactions. These reactions were some of the first well-characterized and well-understood examples of a metabolic pathway.

Some of the landmark experiments that led to our understanding of these reactions involved classic techniques in biochemistry that employed glass reaction flasks into which diced tissue preparations, intermediates, end-products, waste products, and inhibitors could be added independently, in combination, or in sequence. Biochemical activities such as the production of end-products, consumption of intermediates, and rates of enzyme activity were then measured to follow the rates of the reaction. Modern-day biochemists continue to rely on experiments of this design to study metabolism.

Many investigators contributed to research in the late 1800s and early 1900s that unraveled the sequence of reactions that occur during glycolysis. Biochemical studies in the early 1900s demonstrated that preparations of diced animal tissues exhibited the enzymatic ability to carry out oxidation-reduction reactions by transferring hydrogen ions from acids such as citrate, malate, and fumarate to dye molecules that would change color when they were reduced. Albert Szent-Gyorgi carried out some of the earliest studies. When performed under aerobic conditions, his studies established that organic acids could be oxidized to produce carbon dioxide and water, and that inhibitors of these reactions blocked the consumption of oxygen in the reaction.

Another pioneer of cellular respiration was the German biochemist Hans Krebs, who discovered the citric acid cycle, which now bears his name—the Krebs cycle. Krebs proposed that this pathway involved a cycle of enzymatic reactions that were required to oxidize organic acids such as pyruvate. Many of Krebs's experiments involved studying the oxidation of organic acids in slices of flight muscle from pigeons. To support active wing flapping over long distances, these muscle cells depend heavily on oxidative metabolism to produce ATP, so they served as an excellent tissue source for Krebs's research. Krebs also determined that the use of inhibitors such as malonate could block the production of some intermediates in the cycle while causing the accumulation of other intermediates. Using this approach, Krebs was able to determine the sequence of enzyme steps involved in the citric acid cycle by observing which intermediates were consumed and which ones were accumulated following the addition of inhibitors. The importance of Krebs's work was acknowledged when he was awarded the Nobel Prize in 1953, together with Fritz Lipmann who discovered a coenzyme that is an important electron carrier molecule in the electron transport chain.

In the 1920s and 1930s, many scientists worked independently to determine the steps involved in electron transport. In 1961, British biochemist Peter Mitchell proposed a mechanism called chemiosmosis as the process during oxidative phosphorylation that requires a hydrogen ion gradient for the production of ATP. Mitchell carried out most of his research using bacteria. His contribution to cellular respiration was recognized with a Nobel Prize in 1978.

You will use MitochondriaLab to simulate some of the pioneering experiments—similar to those carried out by Szent-Gyorgi, Krebs, Mitchell, and other researchers—that were central to our understanding of cellular respiration. You will add substrates and inhibitors to a reaction flask containing isolated preparations of mitochondria and measure oxygen consumption as a function of the rate of these reactions. The challenge of MitochondriaLab is to carefully analyze your results to develop an understanding of the sequence and control of these reactions.

References

1. Campbell, M. K. *Biochemistry*, 2nd ed. NY: Saunders, 1995.

2. Mathews, C. K., and van Holde, K. E. *Biochemistry*, 2nd ed. Menlo Park, CA: Benjamin/Cummings, 1996.

Introduction
In this laboratory, you will perform simulations of biochemistry experiments designed to study the reactions of aerobic cellular respiration as it occurs in vitro using isolated preparations of mitochondria. By changing reaction conditions through adding substrates and metabolic inhibitors, you will simulate some of the experiments that

were used to determine the metabolic reactions and the sequence of the reactions that occur during the Krebs cycle and the electron transport chain.

Objectives

The purpose of this laboratory is to:

- Demonstrate how studying the effects of intermediates in a reaction can be used to determine the steps in a metabolic pathway and the sequence of each enzymatic step in the pathway.
- Demonstrate how changes in substrates, products, and other reaction parameters can affect the reactions involved in the Krebs cycle, electron transport chain, and oxidative phosphorylation.
- Simulate the effects of substances that inhibit the reactions of aerobic cellular respiration.

Before You Begin: Prerequisites

Before beginning MitochondriaLab you should be familiar with the following concepts:

- The importance and functions of enzymes as biological catalysts, basic principles of metabolic pathways, and mechanisms involved in regulating the catalytic activity of an enzyme (see Campbell, N. A., Reece, J. B., and Mitchell, L. G. *Biology* 5/e, chapter 6).
- The structure and function of the mitochondrion (chapters 7 and 9).
- The reactions of aerobic cellular respiration including glycolysis, the Krebs cycle, the electron transport chain, and oxidative phosphorylation. Be able to describe the primary substrates required, reactions involved, and products generated by each of these reactions (chapter 9).
- The enzymatic steps involved in the Krebs cycle, and the organic acids produced as intermediates during the Krebs cycle (chapter 9).
- Feedback mechanisms involved in the control of cellular respiration (chapter 9).

Assignments

The first screen that appears in MitochondriaLab shows you a biochemistry lab containing all the equipment you will need to perform your experiments. Click on each piece of equipment to learn more about its purpose. The following assignment is designed to help you become familiar with the operation of MitochondriaLab.

Getting to Know MitochondriaLab: Setting Up an Experiment

When you enter MitochondriaLab, you are provided with a reaction flask in which you will perform your reactions. These reaction flasks are a standard piece of glassware in a biochemistry lab because they allow biochemists to carefully control and measure reaction conditions within the flask. To begin each experiment, you will start with an empty reaction flask into which you will add isolated preparations of mitochondria. In real-life applications, these are typically mitochondria that have been isolated from mitochondria-rich animal tissues, such as liver or muscle, using cell fractionation techniques. Notice that you have two racks of test tubes to choose from. One rack of tubes contains seven different substrates that you can add to your flask (ADP, ascorbate/TMPD [tetramethyl-*p*-phenylene diamine], fumarate, glutamate, malate, pyruvate, and succinate). The second rack of tubes contains six different metabolic inhibitors of cellular respiration (antimycin, cyanide, DNP [2,4-dinitrophenol], malonate, oligomycin, and rotenone).

Notice that the chart recorder is connected to an oxygen-measuring electrode that inserts into a port on the left side of the reaction chamber. Substrates and inhibitors will be added to the flask through the port on the right side of the flask. Oxygen consumption in the flask will be determined as a measurement of the progress of reactions in the flask. The chart recorder will measure and plot oxygen concentration in the flask during the time course of your experiments.

1. Begin your first experiment by developing a hypothesis to predict what will happen to oxygen consumption in the reaction flask after the addition of pyruvate. Develop a second hypothesis to predict how oxygen consumption will change in the flask upon the addition of pyruvate and ADP. Recall that pyruvate is an end-product of the reactions of glycolysis. Once you have formulated your hypotheses, setup your simulated experiment as follows:

   a. Click on the To Experiment button in the lower left corner of the screen to enter MitochondriaLab. To begin an experiment you, must first add mitochondria to the reaction flask. Do this by clicking on the Add Mitochondria button located below the reaction flask. A running ledger of the components that you added and the time that each component was added to the flask is shown in the lower right corner of the screen. Notice that the mitochondria have been added to a 2.0 ml suspension that consists of a solution buffered to the proper pH for these reactions. This solution contains ions and other components that are required by mitochondria to undergo respiration.

   The chart recorder at the left of the screen will be used to monitor the rate of cellular respiration by the mitochondria in the flask by plotting oxygen concentration (as a percentage of the starting concentration) on the x-axis against time in minutes on the y-axis. Oxygen concentration in the flask begins at 100%. Notice that the chart recorder is plotting oxygen consumption in blue ink. If desired, you can change the recording speed of the simulation by using the small buttons at the upper left corner of the screen. The default value is 4×.

   b. Allow the experiment to proceed for two minutes. Take note of any changes in oxygen consumption that occur during this time. Is this what you expected? Explain your observations. After two minutes, add pyruvate to the reaction flask by clicking on pyruvate in the substrates box (pyruvate should now be highlighted) and then clicking the Add button. Notice that you added 20 μl of 500 mM pyruvate to the flask. Follow oxygen consumption in the flask for three minutes. What did you observe during this time? Based on what you already know about the reactions of cellular respiration, explain your observations. Was this result what you predicted based on your hypothesis? Why or why not?

c. Add ADP to the flask and follow oxygen consumption. What happened? Was this result what you expected based on your hypothesis? Why or why not. Explain your observations. Explain why oxygen consumption changes so dramatically following the addition of ADP while the addition of pyruvate alone results in slower consumption of oxygen.

d. Click on the View Chart button to see an enlarged view of the plot. Use the slider bar at the right of the plot to move up or down the plot. Notice that the chart recorder labels the time when you added each component, the concentration of pyruvate added, and the concentration of oxygen (measured in nanomoles of oxygen molecules) at the time the pyruvate was added. You can export the values from this plot by clicking on the Add to Notes button in the lower left corner of the screen. A new browser window will appear with the data in your notebook. You can now print your data from this window or save your data to a disk.

e. Click on the Return to Experiment button to return to the experiment in progress. Reset the reaction flask by clicking on the Clean and Reset Chamber button below the flask to prepare the reaction flask for another experiment. Formulate a hypothesis to explain what you think will happen when you add succinate to the flask. Run the experiment by adding mitochondria to the clean flask, allowing the reaction to proceed for one minute, and then adding succinate to the flask. Did the results of this experiment validate your hypothesis? Why or why not? Explain why the addition of succinate produced the observed effect on oxygen consumption.

f. Repeat the procedure in step (e) to perform individual experiments for each of the following substrates: fumarate, ascorbate/TMPD (tetramethyl-*p*-phenylene diamine), malate, and ADP. *Note*: Ascorbate and TMPD are synthetic electron donor molecules that can supply electrons to the electron transport chain. For each experiment, save all data to your notebook so you can compare results. Did the effect on oxygen consumption appear to be the same for each substrate? Describe any obvious differences in oxygen consumption that you observed with the different substrates. Based on what you know about the purpose of each substrate in the reactions of cellular respiration, provide possible explanations for the differences in oxygen consumption that you observed. What differences in oxygen consumption did you observe with the use of ADP as a substrate alone compared with using pyruvate and ADP (step c)? Explain these differences.

The Catalytic Effect of Intermediates and Inhibitors on the Krebs Cycle

One of the important concepts established by many of Krebs's experiments was that the organic acids produced during the citric acid cycle can stimulate oxygen consumption. In addition, Krebs noted that these intermediates are generated in cyclical fashion and that each acid serves as the substrate for the next enzymatic reaction in the cycle. Therefore, once one organic acid is formed, the rate of subsequent reactions is dependent on the production of this organic acid. By following the production and catabolism of organic acids in the pathway it was possible to determine the sequence of enzymatic steps involved in the Krebs cycle.

The use of metabolic inhibitors was an essential aspect of the experiments which determined that the reactions of the Krebs cycle are cyclical. By adding an inhibitor and then measuring the accumulation of an intermediate, it is possible to determine the order or sequence of reactions in the cycle. The following assignment is designed to help you determine the sequence of reactions involved in the Krebs cycle by studying the effects of inhibitors on oxygen consumption.

1. Malonate is an analog of one of the organic acids that is naturally produced during the Krebs cycle. Malonate acts as an inhibitor of the Krebs cycle. Your job is to determine which enzyme in the cycle is inhibited by malonate. Begin an experiment by adding mitochondria and malonate. Allow the reaction to proceed for one minute then add succinate. What happens to oxygen consumption in the experiment? To determine which step in the Krebs cycle is inhibited by malonate, try adding pyruvate to the experiment. Perform similar independent experiments by adding malate, fumarate, and ADP, to mitochondria containing malonate. Try other substrates as well. For each experiment describe what happens to oxygen consumption? What do these result tell you about the sequence of succinate, pyruvate, malate and fumarate relative to where malonate is acting? Based on your results, propose a order of reactions in which each substrate appears and suggest a step in the Krebs cycle that is inhibited by malonate. Consult a textbook if needed to verify your answer and to identify the enzyme inhibited by malonate.

    a. Once you have determined malonate's site of action, explain what would happen to the concentration of each of the following molecules: citrate, isocitrate, α-ketoglutarate, succinate, oxaloacetate, malonate, ADP, and ATP in the experiment if you treated these mitochondria with excess amounts of malonate?

The Effect of Inhibitors on Oxidative Phosphorylation

DNP (2,4-dinitrophenol) and oligomycin are examples of inhibitors called uncoupling agents. Uncouplers inhibit ATP synthesis but allow the other reactions of cell respiration to proceed normally. Develop a hypothesis to predict the effects of each agent on oxygen consumption then test the effects of each uncoupler as follows:

1. Begin an experiment with mitochondria, pyruvate, and ADP. Allow the reaction to proceed for one minute, then add an excess amount of DNP by clicking on DNP and then clicking twice on the Add button. What happens to oxygen consumption? Will the addition of more ADP influence oxygen consumption?

Add ADP and observe what happens. Explain your answers. How might DNP be acting on the reactions of oxidative phosphorylation to cause the change in oxygen consumption that you observed?

    a. What do you think is happening to the concentration of H^+, ADP, and ATP in this experiment? Repeat this experiment using oligomycin. What did you observe? Refer to a biochemistry textbook or discuss your results with your instructor to learn how each uncoupling agent disrupts oxidative phosphorylation.

2. To better understand the coupling effect, and the actions of DNP and oligomycin, perform an experiment with mitochondria, pyruvate, and ADP. Allow this experiment to proceed for two minutes, then add oligomycin. What happened to oxygen consumption during this time? Follow this experiment for two or three minutes. Add ADP to the experiment. What happens? Try adding DNP. What happens to oxygen consumption now? Explain each of these results. Why did ADP and DNP each produce the effects that you saw? Compare results from this and the previous experiment. What can you propose about the possible actions of each of these inhibitors on the reactions of oxidative phosphorylation? Consult a textbook to identify the site of action for each uncoupling agent.

Studying the Efficiency of Oxidative Phosphorylation by Measuring the Rate of Oxygen Consumption

Biochemists who study the efficiency of oxidative phosphorylation in isolated preparations of mitochondria are interested in examining the amount of ATP produced relative to the total amount of energy generated during the oxidation of substrates in reactions of cellular respiration. Recall, however, that ATP production depends on the electron flow produced by oxidizing substrates as well as the concentration of ADP in the mitochondrion. Because oxidative phosphorylation relies on this electron flow as well as ADP, oxidative phosphorylation is said to be "coupled" to respiration (electron flow from substrates to oxygen). One way to measure the efficiency of coupling is to determine what is known as a P/O ratio. The P/O ratio is a measure of the number of ATP molecules that are synthesized as a result of the energy from one pair of electrons that completes the chain by being transferred to one atom of oxygen. A P/O ratio can be calculated by dividing the number (in nmoles) of molecules of ATP synthesized (or ADP consumed) by the number of oxygen atoms (in nmoles) consumed. Depending on the substrates, inhibitors, oxygen concentration, and other parameters, biochemists often categorize oxygen consumption in isolated mitochondria into five energy states. Each state represents different rates of oxygen consumption depending on the components added to the preparation of mitochondria.

Generally, a P/O ratio of approximately 3 (3 moles of ATP) is expected from the oxidation of one mole of NADH. This value is a measure of oxygen consumption during ADP phosphorylation in the ATP-synthesizing step of oxidative phosphorylation catalyzed by ATP synthase. The P/O ratio for any given substrate can be measured. A P/O ratio indicates that a substrate has been oxidized to produce NADH, and the value of the ratio depends on where the electrons from a given substrate or from NADH enter (known as coupling) the electron transport chain.

When substrates donate electrons that enter early in the chain (further from oxidative phosphorylation), these electrons are involved in more transfer reactions. As a result, these electrons contribute more to the H^+ gradient needed for oxidative phosphorylation than electrons coupled to the chain at later stages (thus resulting in a high P/O ratio).

The following exercise is designed to help you understand the effects of different substrates on ADP and oxygen consumption by measuring ATP-to-oxygen ratios in your experiments. You will be measuring the amount of ADP that was phosphorylated to ATP to determine P/O ratios. To calculate a P/O ratio, you need to know the following parameters: (1) the reaction is performed at 25°C, (2) the volume of the reaction flask is 2 ml, and (3) there are 237 nanomoles (nmoles) of molecular oxygen (O_2)/ml of reaction volume at 25°C.

1. For this assignment, we will consider two energy states of mitochondria. State 4 represents the rate of oxygen consumption due to the substrate alone, while state 3 represents oxygen consumption during ADP phosphorylation. Begin an experiment by adding mitochondria, pyruvate, and ADP. Follow this experiment for several minutes.

    a. The mitochondria will go into state 3 when ADP is added. Click on the View Chart button. Note that once all of the ADP has been phosphorylated to ATP this will be labeled on the plot as ADP phosphorylated. At this point, the mitochondria are now in state 3. Recall that the oxygen recorder is plotting molecular oxygen concentration as a percent of the starting concentration (100%). To determine the amount of molecular oxygen consumed while the mitochondria were in state 3, subtract the percent of oxygen present at the time when ADP was fully phosphorylated from the amount of oxygen present when ADP was first added to the experiment. *Note*: For your convenience, use the calculator in MitochondriaLab. Follow the steps below to determine the P/O ratio for pyruvate during state 3.

        (1) To determine the number of nmoles of molecular oxygen present in the reaction flask, multiply 237 nmoles/ml molecular oxygen by 2 ml (the size of the reaction flask). Because P/O values are calculated based on the amount of atomic oxygen consumed, you must now convert nmoles of molecular oxygen into nmoles of atomic oxygen by multiplying nmoles of molecular oxygen in the reaction flask by 2 (recall that there are two oxygen atoms in each molecule of oxygen). This value now represents nmoles of atomic oxygen in the flask.

        (2) To determine nmoles of atomic oxygen consumed in state 3, multiply the percent of molecular oxygen consumed in state 3 by the number of nmoles of atomic oxygen in the flask. (For example, if you observed that 5% of oxygen was consumed during state 3, multiply 0.005 by the number of nmoles of atomic oxygen in the flask that you calculated in step (1)

above.)   This value represents nmoles of atomic oxygen consumed in state 3.

   b. To calculate the P/O ratio in state 3, you must also determine the amount of ADP in nmoles. Recall that you added 20 µl of 100 mM ADP to the reaction flask. Convert this to nmoles first. To determine the P/O ratio, divide nmoles of ADP added by nmoles of atomic oxygen consumed (determined in step 2).

   c. Repeat this experiment using malate, succinate, and ascorbate/TMPD as the substrates. Calculate P/O ratios for each substrate in state 3 and compare these ratios. Refer to a textbook to use diagrams of the Krebs cycle and electron transport chain to trace each substrate to where the substrate itself or electrons from its oxidation enter (are coupled to) the reactions of cellular respiration. Do the ratios make sense to you?

   d. As you learned in the first assignment, Getting to Know MitochondriaLab, ascorbate/TMPD is not a natural substrate combination. From the P/O ratio, you should be able to narrow down the possible location at which TMPD donates electrons to the reactions of cellular respiration.

2. Repeat experiment 1 with glutamate and then fumarate. Are the state 3 ratios different for the two substrates? You will have to look up the glutamate dehydrogenase reaction to determine whether the P/O ratio makes sense, because glutamate is not a direct substrate of the Krebs cycle; it enters the reactions of cellular respiration at a different stage. By what pathway does glutamate donate energy to the reactions of cellular respiration? Trace the path of energy from fumarate. What is the difference between the glutamate and fumarate pathways? What is the rate-limiting step in each reaction pathway?

<u>Group Exercises</u>

Certain poisons act as metabolic inhibitors by binding to molecules in the electron transport chain, thereby preventing electron transport and blocking production of the H^+ gradient required for oxidative phosphorylation. Binding can occur in a reversible or irreversible fashion. Rotenone is a poison that works as an irreversible inhibitor of the electron transport chain. Rotenone is frequently used to selectively kill undesirable species of insects and nonnative fish, such as grass carp, that can cause extensive destruction of plant species in lakes and reservoirs.

The following exercises are designed to help you determine which complex of the electron transport chain is a possible binding site for rotenone and other metabolic inhibitors, as well as examine where other substrates enter the electron transport chain. Before you begin these assignments, you may need to refer to a biochemistry textbook to view a detailed figure of the electron transport chain complexes I, II, III, and IV, and the electron transport molecules involved in each complex. Work together in a group of four students to complete the following assignments.

1. Begin an experiment with mitochondria, ADP, and glutamate. Allow the experiment to proceed for one minute, then add rotenone. *Note*: Keep a record of all your experiments in your lab notebook so you can refer back to your results as you interpret the results from other experiments. What happened to oxygen consumption? Now add ADP to the flask. What happened to oxygen consumption after adding ADP? Is this what you expected? Explain your answer.

   a. Ascorbate can bind to a complex in the electron transport chain and donate electrons to the chain. Use ascorbate to help you pinpoint where rotenone is blocking the chain by repeating the same experiment with mitochondria, ADP, and glutamate; then wait one minute, add rotenone, wait another minute, and then add ADP and ascorbate/TMPD to the experiment. What happened to oxygen consumption? Can you determine which complex of the electron transport chain is bound by ascorbate? Based on these results and knowledge of where ascorbate binds, can you determine which complex (I, II, III, or IV) is likely to be inhibited by rotenone? Continue to the next experiment to pinpoint the site of action of ascorbate and rotenone.

   b. Succinate can also bind to a complex in the electron transport chain and donate electrons to the chain. Use succinate to help you determine where rotenone is blocking the chain by repeating the same experiment as above with mitochondria, ADP, and glutamate; then wait one minute, add rotenone, wait another minute, and then add ADP and succinate to the experiment. What happened to oxygen consumption? Can you determine which complex of the electron transport chain is bound by succinate? Based on these results and knowledge of where succinate binds, can you determine which complex (I, II, III, or IV) is likely to be inhibited by rotenone. Refer to a biochemistry textbook if necessary to determine which complex of the electron transport chain is bound by succinate. Based on your results and knowledge of where succinate binds, which complex (I, II, III, or IV) is likely to be inhibited by rotenone? Continue to the next experiment to pinpoint the site of action of rotenone, ascorbate, and succinate.

   c. Antimycin is another metabolic inhibitor that acts on the electron transport chain. Use antimycin to help you pinpoint where rotenone is blocking the chain by repeating the experiment in step (b) but after adding rotenone, add ADP and antimycin to the experiment. What happened to oxygen consumption? Which complex of the electron transport chain is bound by antimycin? Based on your results and knowledge of where antimycin binds, which complex (I, II, III, or IV) is likely to be inhibited by rotenone? Are all of your results consistent enough to help you pinpoint the binding site of rotenone? Once you have determined which complex is bound by rotenone, discuss your answer with your instructor to find out the specific molecules affected by rotenone, ascorbate, succinate, and antimycin.

2. Repeat experiment (b) above, but after adding succinate and ADP, wait one minute and then add cyanide as a final step in the experiment. Cyanide blocks the

electron transport chain in the same fashion as carbon monoxide (CO), the colorless, odorless gas produced when many organic materials are burned. What happened to oxygen consumption in this experiment? Repeat the same experiment for each of the other experiments you conducted above. For each experiment, what happened to oxygen consumption? Is this what you expected? Why or why not? Can you add substrates to bypass the cyanide block? Try this experiment with all of the substrates available. What did you observe? Explain your results, then determine which complex of the electron transport chain is blocked by cyanide.

3. Based on what you know about the electron transport chain and oxidative phosphorylation, explain how adjusting the pH of your reaction flask can be used to artificially bypass the effects of these inhibitors. What would you add to the reaction flask if you were actually going to perform this experiment? Based on what you know about the actions of oligomycin from earlier experiments, how might this experiment be affected by the addition of oligomycin?

FlyLab

Background

While developing his theory of evolution by natural selection, Charles Darwin was unaware of the molecular basis for evolutionary change and inheritance. Around the same time that Darwin was making his landmark observations that led to his publication, *The Origin of Species,* the Augustinian monk Gregor Johann Mendel was performing experiments with garden peas that dramatically influenced the field of biology. Mendel's interpretation of his experimental results served as the foundation for the discipline known as **genetics**. Genetics is the discipline of biology concerned with the study of heredity and variation. In 1866, Mendel published a classic paper in which he described phenomena in garden peas that laid down the principles for genetic inheritance in living organisms.

Our understanding of genetics in higher organisms such as humans has advanced considerably since Mendel's work. This has occurred, in part, because of significant advances in our understanding of the molecular biology of living cells. The molecular biology revolution that led to our current understanding of genetics was greatly accelerated in 1953 by James Watson and Francis Crick, who revealed the structure of DNA as a double-helical molecule. Considering that a detailed understanding of the structure of DNA, chromosomes and genes, and the events of meiosis were not known during Mendel's time, Mendel's work was a particularly incredible accomplishment. Approximately 40 years passed before the significance of Mendel's insight was realized. Mendel's work gained acceptance as his experiments were replicated and publicized by scientists who performed genetic experiments several years after Mendel died. Mendel's landmark principles on inheritance continue to form the basis for our current understanding of genetics in living organisms. As a modern-day student studying biology, having the benefit of hindsight by studying gene and chromosome structure, and understanding how **gametes** form by meiosis, is a great advantage in developing an understanding of inheritance.

The experimental model for Mendel's work involved performing cross-pollination experiments with a strain of garden peas called *Pisum sativum.* Mendel was interested in studying the inheritance of a number of different characters, or heritable features, of *Pisum.* He considered seven different characters including flower color, flower position, seed color, seed shape, pod color, pod shape, and stem length. Variations of a given character are known as traits. For example, when studying flower color as a character, Mendel traced the inheritance of two traits for flower color, purple flowers and white flowers. Many of the basic genetic principles established by Mendel arose from his observations of the results produced by a simple cross-pollination experiment called a **monohybrid cross**. In a monohybrid cross, individuals with one pair of contrasting traits for a given character are mated. For example, a plant with purple flowers is mated with a plant that has white flowers.

Mendel proposed several basic principles to explain inheritance. One of his first principles was that characters are determined by what he described as "heritable factors" or "units of inheritance." The factors that Mendel described are what we now call genes. Mendel noted that alternative forms of a gene, what we now call **alleles**, are responsible for variations in inherited characters. For example, in *Pisum sativum* there are two alleles for flower color, a white flower allele and a purple flower allele. Mendel also proposed that certain alleles, called **dominant alleles**, are always expressed in the appearance of an organism and that the expression or appearance of other alleles, called **recessive alleles**, was sometimes hidden or masked. Recall that modern-day geneticists frequently use capital letters to indicate dominant genes and lowercase letters to indicate recessive genes. The appearance of a particular trait is referred to as the **phenotype** of an organism while the genetic composition of an organism is known as the organism's **genotype**. Mendel established that an organism typically inherits two alleles for a given character, one from each parent. Using modern genetic terminology, we say that an organism that contains a pair of the same alleles is **homozygous** for a particular trait while an organism that contains two different alleles for a trait is **heterozygous** for that trait. Mendel also postulated that when an organism contains a pair of units, the units separate, or segregate, during gamete formation so that each individual gamete produced receives only one unit of the pair. This postulate became known as Mendel's **law of segregation of alleles**.

One example of a monohybrid cross performed by Mendel involved the inheritance of pea flower color. Mendel began by crossing true-breeding homozygous parents called the **P generation**. He crossed a plant with purple flowers with a plant showing white flowers. The first offspring produced from such a cross, called the F_1 **generation**, all showed purple flowers. The results of this cross were explained when Mendel determined that the F_1 plants were heterozygotes and that these plants showed purple flowers because the purple flower allele was dominant and the white flower allele was recessive. Self-pollination of F_1 plants produced a second generation of plants called the F_2 **generation**, in which the phenotypic ratio of purple-flowered plants to white-flowered plants was approximately 3:1. As he performed monohybrid crosses for several other characters, Mendel discovered that this ratio was characteristic of a cross between heterozygotes. The genotypes and phenotypes that may result from a genetic cross can be predicted by using a Punnett square. You should be familiar with constructing Punnett squares to predict the results of monohybrid and **dihybrid crosses**. Mendel used dihybrid crosses to explain his **law of independent assortment**, in which he postulated that alleles for different characters (for example, pea color and pea shape) segregate into gametes independently of each other. A dihybrid cross between F_1 heterozygotes reveals a phenotypic ratio of approximately 9:3:3:1 in the F_2 generation.

The 3:1 ratio predicted for Mendel's monohybrid cross and the 9:3:3:1 phenotypic ratio predicted for a dihybrid cross are hypothetical expected ratios. When performing a real genetic cross based on Mendelian principles, however, such a cross is subject to random changes and experimental errors that produce chance deviation in the actual phenotypic ratios that one may observe. To accurately evaluate genetic inheritance, it is essential that observed deviation in a phenotypic ratio be determined and compared with predicted ratios.

Chi-square analysis (χ^2) is a statistical method that can be used to evaluate how observed ratios for a given cross compare with predicted ratios. Chi-square analysis considers the chance deviation for an observed ratio, and the sample size of the offspring, and expresses these data as a single value. Based on this value, data are converted into a single probability value (p-value), which is an index of the probability that the observed deviation occurred by random chance alone. Biologists generally agree on a p-value of 0.05 as a standard cutoff value, known as the level of significance, for determining if observed ratios differ significantly from expected ratios. A p-value below 0.05 indicates that it is unlikely that an observed ratio is the result of chance alone. When we predict that data for a particular cross will fit a certain ratio—for example, expecting a 3:1 phenotypic ratio for a monohybrid cross between heterozygotes—this assumption is called a **null hypothesis**. Chi-square analysis is important for determining whether a null hypothesis is an accurate prediction of the results of a cross. Based on a p-value generated by chi-square analysis, a null hypothesis may either be rejected or fail to be rejected. If the level of significance is small ($p < 0.05$), it is unlikely (low probability) that the observed deviation from the expected ratio can be attributed to chance events alone. This means that your hypothesis is probably incorrect and that you need to determine a new ratio based on a different hypothesis. If, however, the level of significance is high ($p > 0.05$), then there is a high probability that the observed deviation from the expected ratio is simply due to random error and chi-square analysis would fail to reject your hypothesis.

It is important to realize that the principles established by Mendel can be easily explained by understanding the chromosomal basis of inheritance. For example, in humans, **somatic cells** contain the **diploid number** ($2n$) of chromosomes, which consists of 46 chromosomes organized as 23 pairs of **homologous chromosomes** (homologues). Chromosomes 1 through 22 are called **autosomes** and the twenty-third pair of chromosomes are called **sex chromosomes**. During meiosis, gamete formation leads to the formation of sex cells that contain a single set of 23 chromosomes—the **haploid number** (n) of chromosomes. Because chromosomes are present as pairs in human cells, genes located on each chromosome are also typically present as pairs. Mendel's law of segregation of alleles is explained by the separation of each pair of homologous chromosomes that occurs during meiosis resulting in each individual gamete receiving only one copy of each chromosome. This law applies to, and is explained by, the separation of **unlinked genes** because these genes are located on different chromosomes. As a result, during meiosis the chromosomes align at the metaphase plate and separate in a random, independent fashion.

An expansion in our knowledge of Mendelian genetics has led to a detailed understanding of different types of inheritance events such as codominance, incomplete dominance, linked inheritance, inheritance of multiple alleles, lethal mutations, and the genetic transmission of a number of different genetic disorders. For many reasons, Mendel's pea plants served as an ideal model organism for him to study. In fact, pea plants allowed Mendel to succeed where others failed. Peas are easy to grow, they cross-fertilize and self-fertilize easily, and they mature quickly. More recently, a number of other organisms have served similar roles as model organisms for scientists who study genetics. Some of these include the flowering plant *Arabidopsis thaliana,* a common inhabitant of home aquariums called the striped zebrafish (*Brachydanio rerio*), many species of mice, and the common fruit fly, *Drosophila melanogaster*.

In particular, *Drosophila* has been one of the most well studied model organisms for learning about genetics and embryo development. These small flies are hardy to grow under lab conditions, and they reproduce easily with a relatively short life cycle compared with vertebrate organisms; hence, crosses can be performed and offspring counted over reasonable intervals of time. Another advantage of *Drosophila* is that the **loci** for many genes on the four chromosomes in the fly's genome have been determined, and a very large number of mutations of the **wild-type** fly have been developed that effect different phenotypes in *Drosophila*. Because of some of these characteristics, *Drosophila* has an important historical place in the field of genetics and continues to be an important model organism for studying genetic inheritance that universally applies to most organisms. Using FlyLab, you will design and carry out experimental crosses using *Drosophila melanogaster*. In the future, you will have the opportunity to use PedigreeLab to study inheritance of different genetic disorders through several generations of humans.

References

1. Klug, W. S., and Cummings, M. R. *Essentials of Genetics*, 2nd ed. Upper Saddle River, NJ: Prentice Hall, 1996.

2. Lindsley, D. and Zimm, G. *The Genome of Drosophila melanogaster.* San Diego: Academic Press, 1992.

3. Orel, V. *Mendel.* New York: Oxford University Press, 1984.

4. Soudek, D. "Gregor Mendel and the People Around Him." *Am. J. Hum. Genet.,* May 1984.

5. Stern, C., and Sherwood, E. R. *The Origin of Genetics: A Mendel Source Book.* San Francisco: W. H. Freeman, 1966.

Introduction

FlyLab will allow you to play the role of a research geneticist. You will use FlyLab to study important introductory principles of genetics by developing hypotheses and designing and conducting matings between fruit flies with different mutations that you have selected. Once you have examined the results of a simulated cross, you can perform a statistical test of your data by chi-square analysis and apply these statistics to accept or reject your hypothesis for the predicted phenotypic ratio of offspring for each cross. With FlyLab, it is possible to study multiple generations of offspring, and perform testcrosses and backcrosses. FlyLab is a very versatile program; it can be used to learn elementary genetic principles such as dominance, recessiveness, and Mendelian ratios, or more complex concepts such as sex-linkage, epistasis, recombination, and genetic mapping.

Objectives

The purpose of this laboratory is to:

- Simulate basic principles of genetic inheritance based on Mendelian genetics by designing and performing crosses between fruit flies.
- Help you understand the relationship between an organism's genotype and its phenotype.
- Demonstrate the importance of statistical analysis to accept or reject a hypothesis.
- Use genetic crosses and recombination data to identify the location of genes on a chromosome by genetic mapping.

Before You Begin: Prerequisites

Before beginning FlyLab you should be familiar with the following concepts:

- Chromosome structure, and the stages of gamete formation by meiosis (see Campbell, N. A., Reece, J. B., and Mitchell, L. G. *Biology* 5/e, chapters 13–15).
- The basic terminology and principles of Mendelian genetics, including complete and incomplete dominance, epistasis, lethal mutations, recombination, autosomal recessive inheritance, autosomal dominant inheritance, and sex-linked inheritance (chapters 14 and 15).
- Predicting the results of monohybrid and dihybrid crosses by constructing a Punnett square (chapter 14).
- How genetic mutations produce changes in phenotype, and beneficial and detrimental results of mutations in a population (chapter 17).

Assignments

To begin an experiment, you must first design the phenotypes for the flies that will be mated. In addition to wild-type flies, 29 different mutations of the common fruit fly, *Drosophila melanogaster*, are included in FlyLab. The 29 mutations are actual known mutations in *Drosophila*. These mutations create phenotypic changes in bristle shape, body color, antennae shape, eye color, eye shape, wing size, wing shape, wing vein structure, and wing angle. For the purposes of the simulation, genetic inheritance in FlyLab follows Mendelian principles of complete dominance. Examples of incomplete dominance are not demonstrated with this simulation. A table of the mutant phenotypes available in FlyLab can be viewed by clicking on the Genetic Abbreviations tab which appears at the top of the FlyLab homepage. When you select a particular phenotype, you are not provided with any information about the dominance or recessiveness of each mutation. FlyLab will select a fly that is homozygous for the particular mutation that you choose, unless a mutation is lethal in the homozygous condition in which case the fly chosen will be heterozygous. Two of your challenges will be to determine the zygosity of each fly in your cross and to determine the effects of each allele by analyzing the offspring from your crosses.

One advantage of FlyLab is that you will have the opportunity to study inheritance in large numbers of offspring. FlyLab will also introduce random experimental deviation to the data as would occur in an actual experiment! As a result, the statistical analysis that you will apply to your data when performing chi-square analysis will provide you with a very accurate and realistic analysis of your data to confirm or refute your hypotheses.

Getting to Know FlyLab: Performing Monohybrid, Dihybrid, and Trihybrid Crosses

1. To begin a cross, you must first select the phenotypes of the flies that you want to mate. Follow the directions below to create a monohybrid cross between a wild-type female fly and a male fly with sepia eyes.

   a. To design a wild-type female fly, click on the Design button below the gray image of the female fly. Click on the button for the Body Color trait on the left side of the Design view. The small button next to the words "Wild Type" should already be selected (bolded). To choose this phenotype, click the Select button below the image of the fly at the bottom of the design screen. Remember that this fly represents a true-breeding parent that is homozygous for wild type alleles. The selected female fly now appears on the screen with a "+" symbol indicating the wild-type phenotype.

   b. To design a male fly with sepia eyes, click on the Design button below the gray image of the male fly. Click on the button for the Eye Color trait on the left side of the Design view. Click on the small button next to the word "Sepia." Note how eye color in this fly compares with the wild-type eye color. Choose this fly by clicking on the Select button below the image of the fly at the bottom of the Design screen. The male fly now appears on the screen with the abbreviation "SE" indicating the sepia eye mutation. This fly is homozygous for the sepia eye allele. These two flies represent the parental generation (P generation) for your cross.

   c. Based on what you know about the principles of Mendelian genetics, predict the phenotypic ratio that you would expect to see for the F_1 offspring of this cross and describe the phenotype of each fly.

   d. To select the number of offspring to create by this mating, click on the popup menu on the left side of the screen and select 10,000 flies. To mate the two flies, click on the Mate button between the two flies. Note the fly images that appear in the box at the bottom of the screen. Scroll up to see the parent flies and down to see the wild type offspring. These offspring are the F_1 generation. Are the phenotypes of the F_1 offspring what you would have predicted for this cross? Why or why not? *Note*: The actual number of F_1 offspring created by FlyLab does not exactly equal the 10,000 offspring that you selected. This difference represents the experimental error introduced by FlyLab.

   e. To save the results of this cross to your lab notes, click on the Results Summary button on the lower left side of the screen. A panel will appear with a summary of the results for this cross. Note the number of offspring, proportion of each phenotype and observed ratios for each observed phenotype. Click the Add to Lab Notes button at the bottom of the panel. Click the OK button to close this panel. To comment on these results in your lab notes, click on the Lab Notes button and move the cursor to the space above the dashed line and type a comment such as, "These are the results of the F1 generation for my first monohybrid cross." Click the Close button to close this panel and return to the Mate screen.

f. To set up a cross between two F_1 offspring to produce an F_2 generation, be sure that you are looking at the two wild-type offspring flies in the box at the bottom of the screen. If not, scroll to the bottom of this box until the word "Offspring" appears in the center of the box. Click the Select button below the female wild-type fly image, then click the Select button below the male wild-type fly image. Note that the two F_1 offspring that you just selected appear at the top of the screen as the flies chosen for your new mating. Click on the Mate button between the two flies. The F_2 generation of flies now appears in the box at the bottom of the screen. Use the scroll buttons to view the phenotypes of the F_2 offspring.

g. Examine the phenotypes of the offspring produced and save the results to your lab notes by clicking on the Results Summary button on the lower left side of the Mate view. Note observed phenotypic ratios of the F_2 offspring. Click the Add to Lab Notes button at the bottom of the panel. Click the OK button to close the panel.

h. To validate or reject a hypothesis, perform a chi-square analysis as follows. Click on the Chi-Square Test button on the lower left side of the screen. To ignore the effects of sex on this cross, click on the Ignore Sex button. Enter a predicted ratio for a hypothesis that you want to test. For example, if you want to test a 4:1 ratio, enter a 4 in the first box under the Hypothesis column and enter a 1 in the second box. To evaluate the effects of sex on this cross, simply type a 4 in each of the first two boxes, and type a 1 in each of the last two boxes. Click the Test Hypothesis button at the bottom of the panel. A new panel will appear with the results of the chi-square analysis. Note the level of significance displayed with a recommendation to either reject or not reject your hypothesis. What was the recommendation from the chi-square test? Was your ratio accepted or rejected? Click the Add to Lab Notes button to add the results of this test to your lab notes. Click OK to close this panel.

i. To examine and edit your lab notes, click on the Lab Notes button in the lower left corner of the screen. Click the cursor below the recommendation line and type the following: "These are my results for the F_2 generation of my first monohybrid cross. These data do not seem to follow a 4:1 ratio." To print your lab notes, you can export this data table as an html file by clicking on the Export button. In a few seconds, a new browser window should appear with a copy of your lab notes. You can now save this file to disk and/or print a copy of your lab notes. Click the Close button at the bottom of the panel to close the panel.

    j.   Repeat the chi-square analysis with a new ratio until you discover a ratio that will not be rejected. What did you discover to be the correct phenotypic ratio for this experiment? Was this what you expected? Why or why not? What do the results of this experiment tell you about the dominance or recessiveness of the sepia allele for eye color?

2. Click on the New Mate button in the lower left corner of the screen to clear your previous cross. Following the procedure described above, perform monohybrid crosses for at least three other characters. For each cross, develop a hypothesis to predict the results of the phenotypes in the F_1 and F_2 generations and perform chi-square analysis to compare your observed ratios with your predicted ratios. For each individual cross, try varying the number of offspring produced. What effect, if any, does this have on the results produced and your ability to perform chi-square analysis on these data? If any of your crosses do not follow an expected pattern of inheritance, provide possible reasons to account for your results.

3. Once you are comfortable with using FlyLab to perform a monohybrid cross, design a dihybrid cross by selecting and crossing an ebony body female fly with a male fly that has the vestigial mutation for wing size. Develop a hypothesis to predict the results of this cross and describe each phenotype that you would expect to see in both the F_1 and F_2 generations of this cross. Analyze the results of each cross by chi-square analysis and save your data to your lab notes as previously described in the assignments for a monohybrid cross. Describe the phenotypes that you observed in both the F_1 and F_2 generations of this cross. How does the observed phenotypic ratio for the F_2 generation compare with your predicted phenotypic ratio? Explain your answer.

4. Use FlyLab to perform a trihybrid cross by designing and crossing a wild-type female fly and a male fly with dumpy wing shape, ebony body color, and shaven bristles. Develop a hypothesis to predict the results of this cross and describe each phenotype that you would expect to see in the F_2 generation of this cross. Perform your cross and evaluate your hypothesis by chi-square analysis. What was the trihybrid phenotypic ratio produced for the F_2 generation?

Testcross

A **testcross** is a valuable way to use a genetic cross to determine the genotype of an organism that shows a dominant phenotype but unknown genotype. For instance, using Mendel's peas, a pea plant with purple flowers as the dominant phenotype could have either a homozygous or a heterozygous genotype. With a testcross, the organism with an unknown genotype for a dominant phenotype is crossed with an organism that is homozygous recessive for the same trait. In the animal- and plant-breeding industries, testcrosses are one way in which the unknown genotype of an organism with a dominant trait can be determined. Perform the following experiment to help you understand how a testcross can be used to determine the genotype of an organism.

1. Design a female fly with brown eye color (keep all other traits as wild-type), and design a male fly with dichaete wing angle (keep all other traits as wild-type). Mate the two flies. Examine the results of this cross and save your data to your lab notes. Add to your data any comments that you would like.

    To determine the genotype of an F_1 wild-type female fly, design a double homozygous recessive male fly with brown eye color and dichaete wing angle, then cross this fly with an F_1 wild-type female fly. Examine the results of this cross and save the results to your lab notes. What was the phenotypic ratio for the offspring resulting from this testcross? Based on this phenotypic ratio, determine whether the F_1 wild-type female male was double homozygous or double heterozygous for the wild-type alleles for eye color and wing angle. Explain your answer. If your answer was double homozygous, describe an expected phenotypic ratio for the offspring produced from a testcross with a double heterozygous fly. If your answer was double heterozygous, describe an expected phenotypic ratio for the offspring produced from a testcross with a homozygous fly.

Lethal Mutations

Five of the mutations in FlyLab are lethal when homozygous. When you select a lethal mutation from the Design view, the fly is made heterozygous for the mutant allele. If you select two lethal mutations that are on the same chromosome (same linkage group, or the "cis" arrangement), then the mutant alleles will be placed on different homologous chromosomes (the "trans" arrangement). Crosses involving lethal mutations *will not* show a deficit in the number of offspring. FlyLab removes the lethal genotypes from among the offspring and "rescales" the probabilities among the surviving genotypes. Hence, the total number of offspring will be the same as for crosses involving only nonlethal mutations. Perform the following crosses to demonstrate how Mendelian ratios can be modified by lethal mutations.

1. Design a cross between two flies with aristapedia mutations for antennae shape. Mate these flies. What phenotypic ratio did you observe in the F_1 generation? What were the phenotypes? Perform an F_1 cross between two flies with the aristapedia phenotype. What phenotypic ratio did you observe in the F_2 generation? How do these ratios and phenotypes explain that the aristapedia mutation functions as a lethal mutation? To convince yourself that the aristapedia allele is lethal in a homozygote compared with a heterozygote, perform a cross between a wild-type fly and a fly with the aristapedia mutation. What results did you obtain with this cross?

2. Design a cross between two flies with curly wing shape and stubble bristles. Develop a hypothesis to predict the phenotypic ratio for the F_1 generation. Mate these flies. What phenotypic ratio did you observe in the F_1 generation? Test your hypothesis by chi-square analysis. Repeat this procedure for an F_1 cross between two flies that express the curly wing and stubble bristle phenotypes. Are the phenotypic ratios that you observed in the F_2 generation consistent with what you would expect for a lethal mutation? Why or why not? Explain your answers.

<u>**Epistasis**</u>

The genetic phenomenon called **epistasis** occurs when the expression of one gene depends on or modifies the expression of another gene. In some cases of epistasis, one gene may completely mask or alter the expression of another gene. Perform the following crosses to study examples of epistasis in *Drosophila*.

1. Design and perform a cross between a female fly with vestigial wing size and a male fly with an incomplete wing vein mutation. Carefully study the phenotype of this male fly to be sure that you understand the effect of the incomplete allele. What did you observe in the F_1 generation? *Note*: It may be helpful to click up and down in this display box to closely compare the phenotypes of the F_1 and P generations. Was this what you expected? Why or why not? Once you have produced an F_1 generation, mate F_1 flies to generate an F_2 generation.

    Study the results of your F_2 generation, then answer the following questions. Which mutation is epistatic? Is the vestigial mutation dominant or recessive? Determine the phenotypic ratio that appeared in the dihybrid F_2 generation, and use chi-square analysis to accept or reject this ratio.

2. Perform another experiment by mating a female fly with the apterous wing size mutation with a male fly with the radius incomplete vein structure mutation. Follow this cross to the F_2 generation. Which mutation is epistatic? Is the apterous wing mutation dominant or recessive?

<u>**Sex Linkage**</u>

For many of the mutations that can be studied using FlyLab, it does not matter which parent carries a mutated allele because these mutations are located on autosomes. Reciprocal crosses produce identical results. When alleles are located on sex chromosomes, however, differences in the sex of the fly carrying a particular allele produce very different results in the phenotypic ratios of the offspring. Sex determination in *Drosophila* follows an X-Y chromosomal system that is similar to sex determination in humans. Female flies are XX and males are XY. Design and perform the following crosses to examine the inheritance of sex linked alleles in *Drosophila*.

1. Cross a female fly with a tan body with a wild-type male. What phenotypes and ratios did you observe in the F_1 generation? Mate two F_1 flies and observe the results of the F_2 generation. Based on what you know about Mendelian genetics, did the F_2 generation demonstrate the phenotypic ratio that you expected? If not, what phenotypic ratio was obtained with this cross?

2. Perform a second experiment by crossing a female fly with the vestigial wing size mutation and a white-eyed male. Describe the phenotypes obtained in the F_2 generation. Examine the phenotypes and sexes of each fly. Is there a sex and phenotype combination that is absent or underrepresented? If so, which one? What does this result tell you about the sex chromosome location of the white eye allele?

Recombination

Mendel's law of independent assortment applies to unlinked alleles, but **linked genes**—genes on the same chromosome—do not assort independently. Yet linked genes are not always inherited together because of **crossing over**. Crossing over, or **homologous recombination**, occurs during prophase of meiosis I when segments of DNA are exchanged between homologous chromosomes. Homologous recombination can produce new and different combinations of alleles in offspring. Offspring with different combinations of phenotypes compared with their parents are called **recombinants**. The frequency of appearance of recombinants in offspring is known as recombination frequency. Recombination frequency represents the frequency of a crossing-over event between the loci for linked alleles. If two alleles for two different traits are located at different positions on the same chromosome (heterozygous loci) and these alleles are far apart on the chromosome, then the probability of a chance exchange, or recombination, of DNA between the two loci is high. Conversely, loci that are closely spaced typically demonstrate a low probability of recombination. Recombination frequencies can be used to develop gene maps, where the relative positions of loci along a chromosome can be established by studying the number of recombinant offspring. For example, if a dihybrid cross for two linked genes yields 15% recombinant offspring, this means that 15% of the offspring were produced by crossing over between the loci for these two genes. A genetic map is displayed as the linear arrangement of genes on a chromosome. Loci are arranged on a map according to map units called centimorgans. One centimorgan is equal to a 1% recombination frequency. In this case, the two loci are separated by approximately 15 centimorgans.

In *Drosophila*, unlike most organisms, it is important to realize that crossing over occurs during gamete formation in female flies only. Because crossing over does not occur in male flies, recombination frequencies will differ when comparing female flies with male flies. Perform the following experiments to help you understand how recombination frequencies can be used to develop genetic maps. In the future, you will have the opportunity to study genetic mapping of chromosomes in more detail using PedigreeLab.

1. To understand how recombination frequencies can be used to determine an approximate map distance between closely linked genes, cross a female fly with the eyeless mutation for eye shape with a male fly with shaven bristles. Both of these genes are located on chromosome IV in *Drosophila*. Testcross one of the F_1 females to a male with both the eyeless and shaven bristle traits. The testcross progeny with both mutations or neither mutation (wild-type) are produced by crossing over in the double heterozygous F_1 female. The percentage of these recombinant phenotypes is an estimate of the map distance between these two genes.

2. To understand how recombination frequencies can be used to determine a genetic map for three alleles, mate a female fly with a black body, purple eyes, and vestigial wing size to a wild-type male. These three alleles are located on chromosome III in Drosophila. Testcross one of the F_1 females to a male with all three mutations. The flies with the least frequent phenotypes should show the

same phenotypes; these complementary flies represent double crossovers. What is the phenotype of these flies? What does this tell you about the position of the purple eye allele compared with the black body and vestigial wing alleles? Sketch a genetic map indicating the relative loci for each of these three alleles, and indicate the approximate map distance between each locus.

Group Assignment

Work in pairs to complete the following assignment. Each pair of students should randomly design at least two separate dihybrid crosses of flies with mutations for two different characters (ideally choose mutations that you have not looked at in previous assignments) and perform matings of these flies. Before designing your flies, refer to the Genetic Abbreviations chart in FlyLab for a description of each mutated phenotype. Or view the different mutations available by selecting a fly, clicking on each of the different phenotypes, and viewing each mutated phenotype until you select one that you would like to follow. Once you have mated these flies, follow offspring to the F_2 generation.

1. For each dihybrid cross, answer the following questions. Perform additional experiments if necessary to answer these questions.
    a. Which of these traits are dominant and which traits are recessive?
    b. Are any of these mutations lethal in a homozygous fly? Which ones?
    c. Are any of the alleles that you followed sex-linked? How do you know this?
    d. Which alleles appear to be inherited on autosomes?
    e. If any of the genes were linked, what is the map distance between these genes?

2. For at least one of your crosses, attempt to perform the cross on paper using a Punnett square to confirm the results obtained by FlyLab.

3. Ask another pair of students to carry out one of the crosses that you designed. Did they get the same results that you did in the F_1 and F_2 generations? Did they develop the same hypotheses to explain the results of this mating as you did? Explain your answer.

4. Once you have completed this exercise, discuss your results with your instructor to determine if your observations and predictions were accurate.

TranslationLab

<u>Background</u>

Genetic information is stored in cells as deoxyribonucleic acid (DNA). In addition to functioning as the hereditary material for living organisms, the information stored in DNA as **genes** is the basis for cell metabolism because all proteins are synthesized from genes. However, DNA is not directly copied into protein. Protein synthesis requires a deciphering of the genetic information stored within DNA whereby the sequence of deoxyribonucleotides in DNA are copied into strands of ribonucleic acid (RNA) during a process called **transcription**. During **translation**, different RNA molecules—specifically messenger RNA (mRNA), transfer RNA (tRNA), and ribosomal RNA (rRNA)—are used to specify the amino acids that are incorporated into a protein. Because it has been well established that this flow of genetic information is universally followed by living cells—including bacteria, yeast, and human cells—this concept is often referred to as the *central dogma of the genetic code*.

The structure of DNA and its role as a genetic material has not always been so clearly understood. In the 1900s, many scientists suggested that proteins were a key component of cell metabolism; however, the association between genes and proteins was not known. During this time, Sir Archibald Garrod and William Bateson observed human patients who demonstrated rare diseases caused by deficiencies in metabolic pathways involving amino acids. Garrod and Bateson followed the patterns of genetic inheritance of these disorders within families and concluded that inherited information controls metabolism in a cell. The terms <u>gene</u> and <u>enzyme</u> were not even used at this time.

In the 1930s, George Beadle and Edward Tatum performed experiments with a bread mold, *Neurospora crassa*, that provided substantial evidence for the relationship between genes and proteins. Beadle and Tatum developed and observed several mutant groups of *Neurospora* that were identified by comparing the ability of these mutants to grow on plates with minimal nutritional medias with that of wild-type *Neurospora*. Beadle and Tatum discovered different mutants that were defective in their ability to synthesize the amino acid arginine. Supplementing these mutants with other amino acids allowed some of these mutants to synthesize arginine. Beadle and Tatum hypothesized that the amino acid supplements that allowed the mutants to synthesize arginine were amino acids that the mutants could not synthesize on their own due to a mutation that caused a loss of enzyme activity. These results led Beadle and Tatum to suggest a *one gene - one enzyme hypothesis*, in which they reasoned that a single gene is important for determining the synthesis of a single enzyme. This hypothesis, however, did not account for how DNA was deciphered by a cell to produce a protein. In the 1940s and early 1950s, several other scientists performed experiments with bacteria and viruses to present strong evidence that genes were made of DNA, but the process by which a protein could be produced from a gene was still unknown.

The discovery of the double-helical structure of DNA by James Watson and Francis Crick in 1953 clearly established that hereditary information in cells is encoded in the nucleotide sequences contained within DNA. Although this landmark discovery represented a significant advance in the history of biological research, a basic question still existed: How was the nucleotide sequence of genes interpreted or decoded by a cell to provide that cell with the instructions for the synthesis of a protein?

A number of different investigators carried out studies to demonstrate that RNA was synthesized from DNA and that RNA performed a central role in protein synthesis. In 1961, Marshall Nirenberg and Heinrich Matthaei used an in vitro cell-free protein-synthesizing system to provide the first evidence for protein-coding sequences of RNA nucleotides. In a cell-free system, organelles such as ribosomes and other factors (including amino acids, tRNAs, an mRNA template, and a number of cofactors required for translation) can be added together to synthesize proteins in a test tube. Extracts of organelles and the molecules required for translation are often isolated by the lysis of bacteria or animal cells that are highly active in protein synthesis. Nirenberg and Matthaei used extracts from bacteria called *Escherichia coli*. The addition of radioactive amino acids to a cell-free extract allows biologists to follow the rate and specific sequences of proteins that are translated in the assay. Because mRNA had only recently been discovered and was not yet easily isolated from cells, Nirenberg and Matthaei synthesized RNA **homopolymers**, single strands of RNA containing only one ribonucleotide in each strand (for example, UUUUUUU, AAAAAAA), by using a bacterial enzyme called polynucleotide phosphorylase. This enzyme does not require a DNA template to synthesize strands of RNA. The homopolymers synthesized by polynucleotide phosphorylase were then added to the cell-free system, and the incorporation of radioactive isotopes into protein was measured. In a cell-frcc system, translation begins at multiple and random sites along a nucleotide sequence. Keep this in mind when completing the assignments for this laboratory.

Although these early experiments did not determine the number of nucleotides required for a codon, Nirenberg and Matthaei concluded that certain RNA sequences coded for specific amino acids. For example, poly A codes for lysine. Subsequent experiments by Nirenberg and others using RNA heteropolymers, combinations of different ribonucleotides, served to further delineate the assignment of specific nucleotide sequences to individual amino acids. In 1964, Marshall Nirenberg and Philip Leder used an increased understanding of the function of tRNA and the function of ribosomes as RNA-binding organelles to establish that the genetic code is interpreted as three-ribonucleotide sequences, called **codons**, that specify only one amino acid. Nirenberg and Leder's experiments also provided a better understanding of how the anticodon portion of a tRNA molecule interacts with a codon by base pairing during translation.

Gobind Khorana performed similar experiments with a cell-free system to which long sequences of RNA molecules consisting of repeating dinucleotides (for example, ACACAC), trinucleotides, and tetranucleotides were added. The results of Khorana's experiments, and the work of many others, served to identify new codons as well as confirm the specificity of many codons that were previously identified. In particular,

Khorana concluded that certain sequences, such as a triplet contained in polymers of GAUA, function as termination signals because they do not code for the incorporation of an amino acid into a peptide.

It was apparent from many of these studies that the genetic code is degenerate or redundant, because although each codon codes for only one amino acid, most amino acids are specified by more than one codon. The *wobble hypothesis* was proposed by Francis Crick to explain how the first two nucleotides of a codon are more important for tRNA binding to an anticodon than the third nucleotide. Modified base-pair rules that occur with U at the third position (for example, U may pair with A or G at the third position of a codon) suggested a rationale for why the number of different tRNAs inside a cell does not need to equal the number of codons that code for amino acids.

In this laboratory you will have the opportunity to simulate many of the early experiments involving cell-free extracts that were essential for deciphering and determining the genetic code. You will investigate how polyribonucleotide sequences that you create can be translated in a cell-free system to produce sequences of amino acids, and you will interpret the results of your experiments to help you learn how the genetic code is deciphered.

References

1. Alberts, B., et al. *Essential Cell Biology*, 1st ed. New York: Garland Publishing, 1998.

2. Beadle, G. W., and Tatum, E. L. "Genetic Control of Biochemical Reactions in *Neurospora*." *Proceedings of the National Academy of Science*, USA 27 (1941): 499-506.

3. Nirenberg, M. W. "The Genetic Code: II." *Scientific American*, March 1963.

Introduction

TranslationLab will allow you to study the importance of the nucleotide sequence of mRNA as the fundamental basis for the genetic code universally deciphered by living cells. You will produce sequences of ribonucleotides that will be translated into protein to simulate the landmark experiments involving cell-free extracts that were essential for interpreting and understanding the genetic code.

Objectives

The purpose of this laboratory is to:
- Study the relationship between the nucleotide sequence of a mRNA molecule and protein synthesis.
- Simulate pioneering experiments that were used to delineate the genetic code.
- Demonstrate how a mutation in the nucleotide sequence of a mRNA molecule results in a change in the amino acid sequence of a protein.

Before You Begin: Prerequisites

Before beginning TranslationLab you should be familiar with the following concepts:

- The structure of a polypeptide (see Campbell, N. A., Reece, J. B., and Mitchell, L. G., *Biology* 5/e, chapter 5).
- The structure and functions of messenger RNA (mRNA), ribosomal RNA (rRNA), and transfer RNA (tRNA) (chapter 17).
- The flow of genetic information in a cell; the major processes involved in transcription and translation (chapter 17).
- Describe how point mutations in a gene can affect the amino acid sequence of a protein (chapters 5, 14, 17).

Assignments

During the late 1950s and early 1960s researchers were able to solve one of the major secrets of life: how genes worked. The problem the researchers were trying to solve was how a linear sequence of four nucleotides (A, G, C, and U) determined the amino acid sequence of proteins, which were made out of up to 20 different amino acids. The following assignments are designed to help you reproduce some of the experiments that the scientists used to figure out how this was accomplished. Your mission, should you choose to accept it, is to crack the genetic code of life.

The Genetic Code

A major step forward in figuring out the code was the discovery by Nirenberg in 1961 that a cell-free extract made from *E. coli* cells could translate RNA added to the extract into proteins. The composition of the newly synthesized proteins could be determined by measuring the incorporation of radioactive amino acids into these proteins as they were translated. In his first experiment he made poly U RNA, using the enzyme polynucleotide phosphorylase, and translated it into a peptide of polyphenylalanine using the cell-free extract. This was definitive proof that RNA could code for the synthesis of proteins and gave the first possible assignment of a nucleotide code to the amino acid it specified.

1. Because having each nucleotide code for only one amino acid would allow for only four different amino acids to be incorporated into a protein, it was obvious to researchers that there had to be a conversion between multiple bases and each amino acid. Would two nucleotides at a time be sufficient to provide enough codons to code for all 20 amino acids? Why or why not? How many amino acids could be coded for by codons containing only two nucleotides? Will three nucleotides per codon work? Why or why not? Explain your answers. To answer these questions using TranslationLab, click the Start Experiment button on the input screen of TranslationLab. For each of the four bottles of ribonucleotides that appear, click on the arrow to select a nucleotide. Do this for two nucleotides initially. Click the Make RNA button to display the sequence of mRNA that you created. Click Add to Notes to create a record of your experiment. To translate this sequence into amino acids, click on the To Translation Mix button. Click Add to Notes to add the amino acid

sequence to your notebook. Continue this process until you are able to answer the questions above.

2. Once it was determined that codons consisted of three-nucleotide sequences, the specificity of each codon could be determined. Use TranslationLab to determine what poly U codes for by performing the following exercise. Click the Start Experiment button on the input screen of TranslationLab. For each of the four bottles of ribonucleotides that appear, click on the arrow to select the uracil (U) nucleotide. Click the Make RNA button to display the poly U sequence of mRNA that you synthesized. Click Add to Notes to create a record of your experiment. To translate this sequence into amino acids, click on the To Translation Mix button. What does poly U code for? Click Add to Notes to add this peptide sequence to the poly U sequence.   Repeat the same procedure to make polynucleotides of each of the other three nucleotides. What amino acids do these polynucleotides code for? Refer to a codon chart (see Campbell, N. A., Reece, J. B., and Mitchell, L. G., *Biology* 5/e, chapter 17). Are the amino acids coded for by the polynucleotides you created consistent with what you would expect based on the codon chart?

3. Although the Nirenberg experiments showed that RNA did determine the amino acids in the protein, they did not show how many bases were used for each codon, whether the codons were overlapping (is the second codon read from the second base of the first codon [overlapping] or from the first base after the last base of the first codon? [no overlap]), or whether there could be bases in between the codons that did not code for anything (AUCGGGAACGGGACAGGGG, for instance, where the G's in between AUC, AAC, and ACA aren't translated into amino acids—just as we use spaces to separate words in a sentence). Khorana developed a means to produce polydinucleotide and later, polytrinucleotide and polytetranucleotide sequences of DNA that could then be transcribed into RNA to be added to the cell-free translation mix.

4. If the code is read two bases at a time, what result would you expect for a polydinucleotide such as AUAUAUAU? Try it and see whether your prediction was correct. From your results can you say whether the code is even or odd? Will you get a different result with UAUAUAUA than you did with AUAUAUAU? This result shows that in these crude extracts translation starts at a random location in the RNA sequence. Translate all possible dinucleotides with TranslationLab. Did you get all of the amino acids? If not, which ones are missing? Did you get any amino acids more than once? Which ones? What does this tell you about the code?

5. From what you have already discovered, what do you think will happen if you use a polytrinucleotide such as AAC? Try it. To help analyze your results, once you have entered the polytrinucleotide sequence, click the Make RNA button and then add this sequence to your notebook by clicking the Add to Notes button. Add the peptide sequence produced from this RNA to your notebook as well. Did you get the result you expected? Explain what

happened. Will ACA or CAA give a different result? From these results can you now tell how many bases there are in a codon? If so, how many are there and how do you know this? Comparing this result with the result from polydinucleotide AC, can you now specify a codon for one of the amino acids incorporated by these templates? If so, which codon and which amino acid go together? By elimination, can you assign another codon–amino acid pair? (*Hint*: Using what you know now, look back at the dinucleotide experiment with AC.) What is it? Try CAC next. Did the results support your codon assignment? Is there evidence here that one of the amino acids must have more than one codon that codes for it? If so, which one? Confirm your results by referring to a codon chart (see Campbell, N. A., Reece, J. B., and Mitchell, L. G., *Biology* 5/e, chapter 17).

6. What do you think will happen if you translate a tetranucleotide? Try translating the tetranucleotide CAAG. Did you get the result you expected? Can you now assign a codon to any of the other amino acids that appeared in problem 5? (Don't worry about any new amino acids that showed up here, just solve the codons for the amino acids in problem 5.) If so what are they? Test your assignment with AACG. Did this confirm your results? Using the above data and any other experiments that are necessary, assign amino acids to all possible codons that do not include G or U, only various combinations of A and C.

7. Now try AU, AAU, and AUU. Did you notice something different this time? What happened, and how would you explain this unusual result? List any new codon assignments that you were able to make from these experiments. Use tetranucleotides to figure out which amino acids go with the codons that can be produced using only A and U. What unusual result did you see with some of the tetranucleotides and what is your explanation for this result?

8. Now try GGG, GGA, GGC, and GGU. What amino acid showed up in all four experiments? Are there any codons shared in common by these four reactions? If not, then what must be true to explain your results? Can you propose a codon or codons for the amino acid that showed up in all four experiments? Do the codons that you've just assigned to this amino acid have anything in common? What is it? Use tetranucleotides to prove that your assignment is correct. Comparing these results with the ones above, can you say whether some positions in the codon are less important than others in specifying which amino acid is coded for?

Altering the Genetic Code: Mutations

Single nucleotide changes (point mutations) in the sequence of a gene can result in changes in the amino acid sequence of a protein produced from the mutated gene. One of the most well studied examples of the effects of a mutation on the sequence of a protein involves the oxygen-transporting protein hemoglobin. A mutation creates an altered form of hemoglobin that produces the genetic disorder called sickle-cell disease (sickle-cell anemia). You will learn more about hemoglobin and the effects of mutations in the hemoglobin gene in HemoglobinLab. The purpose of the following assignment is to demonstrate the effect of a point mutation on the amino acid sequence of a protein.

1. Sickle-cell disease results from a point mutation in the second nucleotide of the codon GAA, which results in a change in the amino acid at position 6 in the hemoglobin protein. Synthesize a mRNA from the trinucleotide sequence GAA. Enter this sequence in your notebook. Translate this mRNA and enter the results in your notebook. Synthesize and translate the trinucleotide GUA and do the same for the trinucleotide GAG. Assign codons to each amino acid produced from the three mRNA sequences. (*Hint*: Consider what you know about the sequences for stop codons from attempting to assign codons for each amino acid.) What amino acid does the codon GAA specify? Which amino acid is incorporated into the sickle-cell hemoglobin molecule when this codon is mutated to GUA? Perform other experiments if necessary to confirm your codon assignments to answer this question.

<u>Group Exercise</u>

Because the genetic code is a universal code in biology, in general the nucleotide sequence of important genes is highly conserved across many different species of organisms. It is very common for 70% or more of the nucleotides in a gene to be conserved among very different organisms. Redundancy in the genetic code allows for small differences in the nucleotide sequence for a given gene without significant variations in the amino acid sequence of a protein. For example, the nucleotide sequence of the gene for insulin, the peptide hormone required for glucose uptake by many body cells, is well conserved (greater than 80% similarity) in many vertebrate species. As a result of this conservation of nucleotide sequence, comparing the peptide sequence for insulin from cows, humans, sheep, dogs, and rats often shows fewer than six or seven differences in amino acid sequence. However, point mutations in certain positions of a codon can create changes that dramatically alter the protein produced. Examples of mutations of this type include frameshift mutations. To help you understand why the nucleotide sequences for important genes are highly conserved, work together in a group of four or five students to complete the following assignment.

Imagine that you have just purified a new protein from the brain of adolescent males that you believe may be responsible for excessive hair-combing behavior. From peptide-sequencing experiments, you have determined that this protein contains the following peptide sequence: Trp-Met-Asp-Gly-Trp-Met. Determining the nucleotide sequence of mRNA that was used to translate this part of the protein will enable you to identify the chromosomal location of this new gene and allow you to isolate and clone this gene. It is known that this peptide sequence is highly conserved among males that demonstrate excessive hair-combing behavior which suggests that this portion of the protein is important for its functions. In addition, a mutant form of this protein has also been discovered that appears to result in the loss of the excessive hair-combing behavior. This mutant sequence arises from a single point mutation in the nucleotide sequence of the normal (wild-type) gene for this protein that creates a frameshift mutation. The peptide sequence from this mutant protein is Met-Tyr-Val-Cys-Met-Tyr. Use TranslationLab to complete the following exercises.

1. Determine the sequence of a mRNA that could be used to translate this peptide.

2. Can you determine another sequence of mRNA that would also code for this peptide? Why or why not? Explain your results.

3. Once you have deciphered the mRNA sequence for the normal protein, introduce changes in this sequence until you have determined the nucleotide sequence that specifies the mutated peptide sequence. Examine this mRNA sequence and identify the codon or codon(s) that were altered to create the mutant peptide.

HemoglobinLab

<u>Background</u>

Virtually every chemical reaction and activity that occurs in a living cell requires proteins. A multitude of different types of proteins perform a wide range of functions that include roles in cell support and shape, cell motility, cell communication, protection against foreign materials, cell reception, cell adhesion, catalytic functions as enzymes, and the transport of molecules. This great diversity of protein functions is a direct result of the structural organization and structural properties of proteins.

The three-dimensional conformation of a protein, also known as **protein structure**, is determined by the arrangement of amino acids that are held together by peptide bonds to form a polypeptide (20 or more amino acids linked together by peptide bonds). The specific sequence or order of amino acids in a polypeptide is known as the **primary structure** of a protein. In many proteins, chemical bonding between amino acids produces proteins with higher-order arrangements known as **secondary** and **tertiary structure**. In addition, for certain proteins—particularly enzymes, structural proteins, and transport proteins—a complete and functional protein consists of multiple polypeptide chains (subunits) that must wrap around each other in an arrangement known as **quaternary structure**. The overall conformation or structure of a given protein provides that protein with the unique structural characteristics that are necessary for the proper functions of that protein. Hence, disruption of protein structure, a process known as protein denaturation, drastically alters the functions of a protein.

One of the most extensively studied examples of the relationship between protein structure and function involves **hemoglobin**, the oxygen-transporting protein in human red blood cells. Adult human red blood cells contain relatively few organelles compared with other body cells. In the most basic sense, human red blood cells are essentially membrane sacs filled with hemoglobin. On average, a single red blood cell contains approximately 250 million hemoglobin molecules! The abundance of hemoglobin in red blood cells and the unique structure of the hemoglobin molecule itself accomplish the primary function of red blood cells: to transport oxygen from the lungs to body cells, tissues, and organs.

In addition to transporting oxygen, the conformation of hemoglobin contributes to the biconcave disk shape of human red blood cells. The shape of these cells provides them with the necessary flexibility to flow through thin-diameter blood vessels, such as capillaries, with a minimal amount of friction.

A single hemoglobin molecule consists of four subunits of a polypeptide known as globin. Globin polypeptides are synthesized from a large family of genes that are highly conserved among many species of vertebrate and invertebrate organisms. This family includes relatives such as myoglobin, an oxygen-storage protein present in most vertebrates. Several different globin polypeptides are used to transport oxygen in red blood cells, including some that are used only during fetal development. Adult human red blood cells contain hemoglobin molecules, which involve two alpha globin

(α-globin) subunits and two beta globin (β-globin) subunits that wrap around each other. In the center of each globin subunit is a single iron-containing organic ring known as the heme group. One oxygen molecule can bind to the iron atom in each heme group; therefore, each hemoglobin molecule can bind to and transport a maximum of four molecules of oxygen.

The oxygen-carrying capability of hemoglobin depends on the electron configuration of the iron atom. Within the heme group, the iron atom exists as a transition metal in a divalent state called ferrous iron (Fe^{+2}). The charged nature of an iron ion is essential for oxygen to bind to the heme group. In addition, oxygen binding to the heme group requires a change in the oxidation state of iron that allows the iron to bind oxygen without oxidizing the oxygen or the iron atom itself. This is accomplished by complexing the iron to four nitrogen atoms in the porphyrin ring and one amino acid in the globin subunit, and surrounding the ring with a cluster of hydrophobic amino acids in each globin subunit, thereby creating a hydrophobic pocket around the heme group. This configuration holds the iron atom at a displaced position above the plane of the heme group, which is the ideal conformation for binding and holding oxygen.

In addition, hemoglobin is an efficient carrier of oxygen because the globin chains exhibit an interaction known as cooperativity. In cooperativity, binding of a single oxygen molecule to one heme group results in an increased binding affinity for oxygen to the three other heme groups. Cooperativity occurs because the binding of one oxygen molecule to one heme group creates a shift in the conformation of each globin chain that produces a change in the overall quaternary structure of the entire hemoglobin molecule. This transition in conformation creates a molecule that favors the binding of additional oxygen molecules. The conformation change that occurs during cooperativity is due to the bonds or contacts of specific amino acids that connect the alpha and beta chains to each other and allow these chains to interact with each other.

In 1949, Linus Pauling provided biologists with a significant insight into the molecular basis for **sickle-cell disease** (sickle-cell anemia) by using **gel electrophoresis** to demonstrate that hemoglobin molecules isolated from normal patients and from those with sickle-cell disease differed in their rate of migration. This observation led Pauling to suggest that a difference in amino acid sequence accounted for the migration differences of these proteins. Using peptide sequencing techniques, Vernon Ingram subsequently demonstrated that the differences between normal hemoglobin and sickle-cell hemoglobin are due to an amino acid difference in the primary structure of the two proteins. This amino acid difference occurs because of a point mutation in one of the globin genes.

The most common mutation in a globin gene of an individual with sickle-cell disease involves a substitution in the codon that codes for the amino acid glutamic acid at position 6 in the β-globin polypeptide. Recall that single-nucleotide changes in the DNA sequence of a gene are known as **point mutations**. Point mutations in a gene sequence can result in the synthesis (transcription) of messenger RNA (mRNA) molecules with an altered base sequence. Depending on the location of a mutation within a codon, a mutation may or may not affect the protein coded for by a particular mRNA. The altered codon in sickle-cell disease results in the substitution of valine for glutamic acid at position 6. This disruption in the primary structure of the globin polypeptide occurs in a location on the globin subunit that is necessary for the proper folding of hemoglobin into its three-dimensional conformation that is essential for it to function as an oxygen-transport protein. As a direct result of this change in hemoglobin structure, sickle-cell hemoglobin binds oxygen with a much lower affinity than normal hemoglobin. In addition, red blood cells in the sickle-cell patient lose their characteristic biconcave disk shape and assume an irregular, elongated sickled configuration that greatly diminishes their movement through blood vessels. Sickled red blood cells frequently clump together and block blood flow through the capillaries. Because of changes both in hemoglobin and red blood cell structure, the sickle-cell patient suffers from decreased oxygen delivery to body organs and a variety of other painful conditions related to poor circulation of red blood cells and inadequate oxygen content within the body. The study of hemoglobin biochemistry and sickle-cell disease has provided biologists with an invaluable understanding of the molecular basis for disease.

In this laboratory you will have the opportunity to study the importance of amino acid sequence to the structure of the normal hemoglobin protein and normal human red blood cells. You will also investigate the connections between the nucleotide sequence, the physical properties of the hemoglobin polypeptide, the structure of red blood cells, and the physiological effects of a hemoglobin mutation.

References

1. Ingram, V. M. Gene mutations in human hemoglobin: The chemical difference between normal and sickle-cell hemoglobin. *Nature* 180 (1957).

2. Dickerson, R. E., and Geis, I. *Hemoglobin: Structure, Function, Evolution, and Pathology*. Menlo Park, CA: Benjamin/Cummings, 1983.

3. Pauling, L., Itano, H. A., Singer, S. J., and Wells, I. C. Sickle-cell anemia, a molecular disease. *Science* 110 (1949).

4. Klug, W. S., and Cummings, M. R. *Essentials of Genetics*, 2nd ed. Upper Saddle River, NJ: Prentice Hall, 1996.

Introduction

HemoglobinLab will allow you to study the biochemistry of hemoglobin and the relationship of hemoglobin structure and function to the structure and function of human red blood cells. You can be a biochemist who will use techniques such as gel electrophoresis, peptide sequencing, and computer modeling to study hemoglobin structure. Or you can be a molecular biologist and study the relationship between DNA sequence, polypeptide sequence, and hemoglobin structure.

Objectives

The purpose of this laboratory is to:

- Study the structure and function of hemoglobin, the oxygen-transporting protein in human red blood cells.
- Examine the effects of mutations in the globin gene on hemoglobin structure and human red blood cell structure.
- Demonstrate biochemical techniques that can be used to study protein structure.
- Investigate the process of translation by studying the effects of DNA sequence change on the peptide produced using computer modeling.

Before You Begin: Prerequisites

Before beginning HemoglobinLab you should be familiar with the following concepts:

- The relationship between protein structure and function; the levels of protein structure (see Campbell, N. A. Reece, J. B., and Mitchell, L. G., *Biology* 5/e, chapter 5).
- The use of *TranslateIT* to study transcription and translation.
- How point mutations in a gene can affect protein structure; how a single amino acid substitution mutation in the globin gene results in sickle-cell disease (chapters 5, 14, 17).
- The composition of human blood (chapter 42).
- The use of gel electrophoresis to separate macromolecules (chapter 20).

Assignments

These assignments will enable you to study the effects of mutations in the globin gene for 17 patients and learn how these mutations affect the health of each patient. The following assignment is designed to help you become familiar with the operation of HemoglobinLab by studying sickle-cell disease.

Getting to Know HemoglobinLab: Sickle-Cell Disease

Chills, fever, headache, and vomiting are but a few symptoms of the disease called malaria. Malaria is caused by a protozoan, *Plasmodium vivax,* that lives in tropical countries and reproduces in a mosquito called the *Anopheles* mosquito. *Plasmodium* is transmitted to humans when an infected mosquito bites a human and sporozoites, an infectious stage of *Plasmodium,* enter the human bloodstream and travel to the liver. Sporozoites reproduce in liver cells and release progeny called merozoites into the bloodstream which infect red blood cells, reproduce, and rupture these cells to infect more red blood cells. Individuals who are heterozygous for sickle-cell disease have a higher resistance to malaria than wild type individuals. This resistance occurs because the fragile structure of sickled red blood cells interrupts the life cycle of *Plasmodium.*

1. Select the Blood Samples view on the input screen for HemoglobinLab. Scroll down the Select Case list and choose patient Miriam Dembele. Read Miriam's case history. Note that her case history is consistent with increased resistance to malaria. Compare Miriam's blood sample with the healthy control sample. Are there any obvious differences? Select the Microscope view and make note of any obvious differences in red blood cell structure. Do any of the red blood cells show phenotypic characteristics of sickle-cell disease? If so, approximately what percentage of her cells show these characteristics?

2. Select the Gel Electrophoresis view to examine the electrophoretic migration pattern for the β-globin subunits of Miriam's hemoglobin as compared with a control sample from a healthy patient. Is the migration pattern of Miriam's hemoglobin indicative of a mutation in one of her globin genes? Is Miriam homozygous or heterozygous for this mutation? Explain your answer.

3. Select the Peptide Sequence view. Click the Find Difference button to identify the amino acid change in Miriam's hemoglobin compared with the normal control hemoglobin. Differences in the amino acid sequence of Miriam's hemoglobin protein compared with the normal protein will align at the far left of the screen. Which amino acid has been substituted for in Miriam's gene? Note the position of this amino acid change. This will be important for identifying the position of the nucleotide change in the globin gene.

4. Select the Edit DNA Sequence view. Miriam's globin gene sequence appears, compared with the normal, wild-type globin gene sequence. First, you will need to locate the DNA sequence with the triplet ATG that indicates the position of the start codon that would appear on globin mRNA produced by transcription of this gene. You can do this either by scanning the gene by clicking on the double arrows or (more easily) by typing ATG in the Search window and hitting the return key. This will take you to the ATG with the A in nucleotide 87 outlined with a red box. Click on the Bracket Codons button to outline codons beginning at the ATG. Use the single arrow to advance to codon 6 (nucleotides 102–104).

Click on nucleotide 103—it should now be outlined with a red box—and change the A to a T. Click the Translate button to see a comparison of your custom-mutated protein to Miriam's protein sequence and the normal protein. Is this mutation consistent with what you know about the most common mutation that causes sickle-cell disease? Refer to a codon chart by clicking on the Genetic Code button at the top of the HemoglobinLab homepage and identify the normal codon and the mutated codon that you changed to simulate the amino acid change in Miriam's hemoglobin.   (Remember that HemoglobinLab is showing you DNA sequence, you will need to convert this sequence into a sense strand of DNA and then into a sequence of mRNA.)

Gel Electrophoresis of Hemoglobin

The initial diagnosis of a mutation in a hemoglobin gene often involves the interpretation of a patient's clinical symptoms, patient histories, and the results of biochemical tests such as gel electrophoresis and DNA sequencing. The following assignments are designed to illustrate the importance of gel electrophoresis as a technique that can be used to study protein structure.

1. A number of mutations in the hemoglobin protein result in a mutant protein that demonstrates a faster electrophoretic migration pattern on a gel than the normal protein. Pretend you are a biochemist who is interested in identifying

hemoglobin mutations of this kind. You have available to you 17 patients who have donated blood samples to your lab, and your laboratory technician has run gels on the hemoglobin samples from these patients. It is now your job to interpret these gels to identify which patients may contain the mutant forms of hemoglobin that you are interested in learning more about.

Select the Gel Electrophoresis view, and examine the electrophoresis pattern for the hemoglobin molecules from each of the patients by clicking on each patient's name. Which one of the patients has hemoglobin molecules that show a faster electrophoretic migration pattern than the control molecules?

Select the Peptide Sequence view and click the Find Difference button to identify the altered amino acid sequence for this patient. What is the mutation that appears? Is this mutation at the N-terminus or C-terminus of the globin polypeptide?

Read this patient's case history. This patient's mother has symptoms indicative of a hemoglobin mutation, but the patient's father appears normal. Is the gel electrophoresis pattern that you observed for this patient's hemoglobin consistent with his or her family history? Explain your answer. After you meet with this patient and discuss your interest in his or her blood, the patient tells you that his or her mother was born and raised in Hiroshima, Japan. Because you are a well-rounded person with a strong knowledge of world history, what might you consider to be the cause of this patient's hemoglobin mutation?

2. Certain mutations in the β-globin gene result in altered amino acid sequences in the hemoglobin molecule that produce a protein with an *increased* affinity for binding to oxygen. One example of such a mutation produces a molecule called hemoglobin Yakima. Yakima involves an amino acid substitution mutation at position 99, where aspartic acid is replaced by histidine. Individuals with these mutant forms of hemoglobin often show redder-than-average complexions. Select the Blood Samples view and scroll through the patient case histories searching for the patient whose complexion matches this description.

Once you think you have found this patient, select the Microscope view and evaluate the patient's red blood cells for any obvious defects. Select the Gel Electrophoresis view. Does the migration pattern of this patient's hemoglobin indicate a mutation in the protein? If your answer is yes, does this patient appear to be homozygous or heterozygous for this mutation?

Select the Peptide Sequence view and click the Find Difference button to identify the altered amino acid sequence for this patient. What is the mutation that appears? Is this mutation indicative of hemoglobin Yakima? Provide reasons why you think this mutation may increase the affinity of oxygen for binding to hemoglobin Yakima.

<u>**Peptide Sequence Analysis of Hemoglobin**</u>

As powerful as gel electrophoresis is as a technique, changes in the amino acid sequence of a protein can be definitively determined only by analyzing DNA and peptide sequences. The following assignments will help you understand how these techniques can be applied to study protein structure.

1. Select the Blood Samples view and click on patient Rhonda Emolina. Compare the color of Rhonda's blood with that of the healthy control blood. Is the color of Rhonda's blood consistent with the conditions described in her patient history?

   To determine the cause of Rhonda's anemia (abnormally low number of red blood cells), select the Gel Electrophoresis view. Does Rhonda's hemoglobin migrate differently than the healthy control sample? Because the results of this gel electrophoresis experiment are inconclusive in determining whether the cause of her anemia is due to a hemoglobin mutation or another problem such as an iron deficiency, more information about the sequence of Rhonda's hemoglobin protein needs to be considered.

   Select the Peptide Sequence view and click on the Find Difference button to determine whether Rhonda may contain a mutated version of hemoglobin. Is there a mutation? If so at which position and what amino acid is changed?

2. Many invariant amino acid positions around the heme group have been identified in vertebrate and invertebrate hemoglobins. Changes in these invariant amino acids, many of which form the hydrophobic pocket around the heme group, result in a number of very serious blood diseases including a wide variety of anemias. Anemia is a condition that involves an abnormally low oxygen-carrying capability of the blood. Anemias can occur if the number of red blood cells in an individual is low. Anemias are also caused by the production of abnormal hemoglobin molecules (such as sickle-cell anemia, and thalassemia) and inadequate hemoglobin content in red blood cells. Common symptoms of anemias are decreased blood oxygen content, fatigue, shortness of breath, pale skin, and cool body temperature.

   One of the invariant amino acids in β-globin is mutated in the form of hemoglobin called hemoglobin Hammersmith. This invariant amino acid is located at position 43 in the β-globin polypeptide. This mutation results in an unstable hemoglobin that cannot hold and position the heme group in the proper orientation for oxygen binding. Select the Peptide Sequence view and compare the hemoglobin sequence for each female patient with the control sequence by clicking on the Find Difference button until you identify the female patient with hemoglobin Hammersmith. (Be sure to begin your search by starting at the first amino acid in the protein).

   Which amino acid is substituted for in this patient? Use the Edit DNA Sequence view to identify the codon for this amino acid. Alter this codon by changing positions in the codon until you have recreated the Hammersmith mutation. Refer to the codon chart in HemoglobinLab to confirm the mutation that you created.

Group Exercise

In the previous exercises we have considered alterations in hemoglobin structure created by base-pair substitution mutations in the globin gene. Working together in a group of four or five students complete the following assignment which considers other types of mutations.

1. Select patient Juan Rodriquez. Click on the Blood Samples view and read his history. Note any differences in the color of Juan's blood compared with the control cells. Select the Microscope view and note the appearance of Juan's red blood cells compared with the control cells. Select the Gel Electrophoresis view and describe the electrophoretic pattern of Juan's hemoglobin. Based on this electrophoretic pattern, develop at least two hypotheses that could explain this observation.

2. To determine if any of your hypotheses are correct, select the Peptide Sequence view. Click on the Find Differences button. Compare the amino acid sequence of Juan's hemoglobin with the control sequence.

   a. Once you have determined the first amino acid difference, refer to the codon chart in HemoglobinLab and identify the codon for the amino acid on the normal protein sequence. Select the Edit DNA Sequence view to modify this codon and recreate this mutation.

   b. Return to the Peptide Sequence view. Use the double arrows to examine the rest of Juan's hemoglobin. Record all differences in amino acid sequence that you observe. Do the results of this examination confirm or refute your hypothesis? If necessary, formulate a new hypothesis to account for this observation.

   c. Return to the Edit DNA Sequence view. Based on your hypothesis, alter the codon sequence of Juan's globin gene until you have identified the nucleotide change(s) that have occurred in Juan's globin gene.

   d. What is an aplastic crisis? What are common causes of aplastic anemia? Consider Juan's patient history. How might his aplastic crisis have resulted in the hemoglobin mutation that he has? If necessary, refer to an anatomy and physiology textbook in your library to answer this question.

PedigreeLab

<u>Background</u>

Over a century ago, Gregor Mendel established many of the basic laws of inheritance. In the years following Mendel's discoveries, his work was confirmed and extended by many geneticists. During this time an increased knowledge about DNA structure, and the events of mitosis and meiosis, led to a greater understanding of the chromosomal basis for inheritance which explained many of Mendel's hypotheses. For example, Mendel demonstrated that genes on different chromosomes, or unlinked genes, are inherited by independent assortment. However, when following the inheritance of **linked genes**—genes on the same chromosome—gametes are often produced during meiosis that contain chromosomes with different combinations of alleles compared to the parent chromosomes.

One of the key events of meiosis that has a profound effect on genetic inheritance occurs during prophase of meiosis I. In this stage, homologous chromosomes pair together, in a process called **synapsis**, to form a tetrad. Once paired, a reciprocal exchange or swapping of DNA occurs between segments of DNA on the nonsister chromatids of each homologue. This exchange of DNA is called **crossing over** or **homologous recombination**. Crossing over results in the mixing of alleles between the two homologues. After a single crossover event, each chromosome contains DNA obtained from its homologue. Crossing over can produce new combinations of linked alleles that were not present on the parental chromosomes thus crossing over is one of the important cellular events responsible for the tremendous genetic variation in offspring that arise from gamete formation by meiosis and subsequent sexual reproduction. Offspring that arise from recombinant gametes produced by crossing over are called **recombinants** because they express phenotypes that are different from the parents. Geneticists refer to recombination frequency as a measure of the number of crossover events that occur between two **loci** on a chromosome. On average, one to four crossover events occur between most homologous chromosomes during meiosis in humans. Unfortunately, crossover events can occasionally occur between nonhomologous chromosomes. These events are called **translocations**. Translocations are the underlying molecular cause of certain types of human genetic disorders. For example, one type of leukemia called chronic myelogenous leukemia arises from a translocation of DNA between chromosome 21 and chromosome 14.

The position of DNA exchange between nonsister chromatids during crossing over is called the **chiasma** (pl., chiasmata). Chiasmata can be visualized with an electron microscope as X-shaped configurations where DNA from the homologues overlaps and a complex of enzymes carries out the reactions that result in DNA exchange. Because chiasmata formation and crossing over is, in part, a chance event, the probability of a single crossover event between two loci on a single chromosome is directly proportional to the distance between the two loci. For example, alleles that are far apart on a chromosome are more likely to be involved in crossing over, compared to two closely space loci, because the probability that chiasmata will form between the distant loci is high. Genes that are closer together on the same chromosome will tend to be inherited together because the probability of chiasma formation and crossing over between closely spaced loci is relatively low. Hence, the distance between genes on a chromosome can be estimated by studying the results of genetic linkage studies based on recombination frequencies. Recall that recombinants that result from crossing over between two linked genes can be counted by studying the offspring of a cross between an individual who is a heterozygote for both genes with an individual that is (double) homozygote for both genes. Using recombination frequency values from such a cross, the relative location of the two genes on the same chromosome can be used to develop a genetic map of a chromosome or linkage map, which is an approximate measure of the linear arrangement of loci on a chromosome. Geneticists refer to all of the genes that are connected together on the same chromosome as a linkage group.

In a genetic map, the distance between loci is indicated in map units called centimorgans (cM). One centimorgan is equivalent to a 1% recombination frequency. One centimorgan is equal to approximately 1 megabase (Mb) or one million base pairs of DNA. For two closely spaced loci, maps units are equivalent to the percent of recombinants produced; however, for loci that are far apart, double exchanges of DNA between the two loci (double-crossovers) maintain the allele combinations of the parent effectively reducing the number of recombinants. In this case, map distance is underestimated by comparison of recombination frequencies alone. Map distance must then be measured by multiplying the total number of recombinants by 100 then dividing this value by the total number of offspring.

One of the first geneticists to study recombination, and suggest the use of recombination data as a way to map genes and chromosomes, was Sir Thomas Morgan. In the early 1900s, Morgan was studying inheritance in *Drosophila melanogaster* when he noticed that white-eyed male flies would occasionally appear in a line of flies with wild-type eye color. Morgan hypothesized that the molecular explanation for genetic linkage was the location of a gene on a chromosome, and that the development of recombinant offspring was the result of crossing over during meiosis. A student of Thomas Morgan's, Alfred H. Sturtevant, extended Morgan's observations when he discovered that recombination frequencies between linked genes are additive. For example, when studying recombination frequencies between three linked genes, the location and order of these genes on a chromosome can be arranged in a linear fashion according to the recombination frequency between each locus. The total recombination frequencies between all three loci represents the total length (in centimorgans) occupied by this linkage group.

It is obviously not possible to study the inheritance of linked traits in humans by purposely designing crosses between individuals and carrying out experiments analyzing large numbers of offspring. How then can the inheritance of human genetic disorders be studied? And how can human gene maps and chromosome maps be developed? One technique that is particularly useful for determining the mode of inheritance for a human disorder involves developing a **pedigree**. A pedigree represents a family tree of genetics. In a pedigree, individuals in a family are shown according to their phenotype for a given trait. When a pedigree is developed, symbols are used to signify each family member. A female is represented as a circle and a male is represented as a square. The symbol for an individual who expresses the phenotype for a certain genetic trait is shaded with a particular color. Parents are connected together by a horizontal line while vertical lines from the parents indicate the offspring produced from these parents. Geneticists refer to offspring as *sibs*, an abbreviation for siblings. Sibs are connected to each other by a horizontal line (sibship line). Sibs are shown from left to right in the order in which they are born. In some pedigrees the sibs may be numbered.

By carefully studying the appearance of a particular phenotype in several generations of individuals, especially in a large pedigree, it is often possible to determine the mode of inheritance of that trait. For example, it may be possible to establish whether the trait is inherited by autosomal recessive, autosomal dominant, or sex-linked inheritance. Each of the pedigrees in PedigreeLab traces the inheritance of a single trait. There are limitations to studying inheritance by pedigree analysis. Typically pedigrees tend to follow small numbers of individuals and they may often lack information about certain family members. Although a pedigree is a useful technique for evaluating the mode of transmission for a trait, it is not as reliable as carrying out crosses that produce large numbers of offspring. Also, pedigree analysis of phenotypes alone cannot be used to develop chromosome maps.

Relatively recent advances in molecular biology techniques, combined with pedigree analysis, make it possible to map genes to chromosomes with a high degree of accuracy. One important technique that is widely used to identify the location of a gene is called **restriction fragment length polymorphism (RFLP) analysis**. This technique is based on the fact that specific breaks or cuts in the nucleotide sequence of a piece of DNA can be made using restriction enzymes. Restriction enzymes cut DNA at specific nucleotide sequences, usually from four to six base pairs in length. For example, the restriction enzyme *Eco*RI recognizes and cuts between the G and A in both strands of DNA containing the sequence GAATTC. A large number of different restriction enzymes with known recognition sequences are readily available commercially. Because, with the exception of identical twins, no two individuals have the exact same nucleotide sequence of DNA in their genome, cutting the DNA from two different individuals with the same enzymes will yield different patterns of variable-length DNA fragments, or RFLPs. An individual's RFLP pattern represents a unique "DNA fingerprint" due to allelic variations in that person's genome. These RFLP patterns are inherited by offspring, and individuals can be assigned a genotype depending on whether they are homozygous or heterozygous for restriction enzyme sites on a particular RFLP.

In humans, RFLPs can be used to test for the presence or absence of a particular allele, especially if the nucleotide sequence of the gene in question is known. If an individual contains a mutation that affects the DNA sequence of a restriction site (either by eliminating a site or by creating a new cutting site), then cutting this DNA with a restriction enzyme will produce different restriction fragments compared with DNA containing the wild-type allele. In this technique, chromosomal DNA is isolated from a tissue sample such as skin, hair, or white blood cells. The DNA is then cut into fragments (RFLPs) with one or more restriction enzymes. These fragments are separated by **gel electrophoresis**. Following electrophoresis, the RFLPs are transferred to a nylon filter that binds the DNA. The filter is then mixed together with a radioactive DNA probe that hybridizes to complementary sequences in RFLPs on the filter. This procedure is called **Southern blotting**. Fragments bound to the probe are visualized by exposing the Southern blot to X-ray film. Exposing the film to the radioactive probe produces dark bands on the film. The exposed film is called an autoradiogram.

What if the nucleotide sequence for a gene is not known? While RFLP analysis is useful for determining genotypes by searching for the presence or absence of a known allele, RFLP analysis must be combined with other techniques when the nucleotide sequence of a gene is not known. How can geneticists find a gene that is responsible for causing a rare genetic disorder? It is possible to look for linkage between the gene causing the disorder and another gene only if other genes close to the mutant gene have already been discovered. Quite frequently, genes near the mutant gene of interest are not known. In these cases, one way to study linkage to map a mutant gene depends on sequences of chromosomal DNA called markers. A marker is typically a sequence that is known to be located at a specific location on a chromosome. Sometime markers include known genes that have been already been identified, but more commonly, markers are short non-protein-coding sequences of DNA. One very valuable group of human chromosome markers are called microsatellites. Microsatellites are usually tandem repeats of dinucleotides (e.g., CACACA) that are highly repeated from individual to individual. Because they are scattered throughout the genome, microsatellites are excellent markers for determining and mapping how close an unknown mutant gene may be to a microsatellite marker on many different human chromosomes. A marker can be detected by hybridizing short pieces of radioactive DNA, called probes, that are complementary to the nucleotide sequence of the marker.

RFLP analysis using markers has proven to be a very powerful technique for mapping chromosomes. Piecing together overlapping RFLPs and other markers has enabled biologists to develop extensive linkage maps of chromosomes in the human genome. In excess of 5000 markers representing different loci in the genome have been identified and are available for use by scientists who are searching for genes that cause inherited disorders in humans. With so many markers available, radioactive probes for many different markers can be hybridized to RFLP fragments from individuals who have a certain disease. RFLP patterns from many individuals with a disease can be compared with each other and compared against RFLP patterns for normal, healthy individuals to search for markers that are linked to the same chromosome as the defective gene.

Phenotypic information gathered from pedigrees combined with RFLP analysis provides very powerful tools that can be applied by geneticists and molecular biologists who are attempting to identify and locate mutant alleles that cause human diseases. The gene that causes cystic fibrosis was one of the first genes for a human genetic disorder that was mapped using RFLP techniques. Cystic fibrosis is an autosomal recessive disease in which individuals suffer from severe respiratory problems, among other symptoms. The cystic fibrosis gene, located on chromosome 7, codes for a protein that normally functions to pump chloride ions out of many body cells. Once the chromosome containing the cystic fibrosis gene was identified, it was possible to study recombination between the cystic fibrosis gene and linked markers to pinpoint the exact locus of the cystic fibrosis gene. Refer to Figure 1 on page 17 for an example of RFLP patterns generated for an individual who is heterozygous for an autosomal recessive condition.

Statistical analysis can be applied to the recombination data generated by RFLP studies to determine the probability that a marker and trait are linked. Based on crossing over, recombination data between a marker and a trait can be assigned an LOD score, which is a statistical comparison of the probability that RFLP data is due to linked loci and the probability that RFLP data is due to unlinked loci. A standard LOD score of 3.0 has been established as the threshold value for linkage. A LOD of 3.0 is equal to approximately a 20:1 ratio that favors linkage. A LOD value of 3.0 or higher is good evidence that two loci are linked, and the recombination frequency can then be used assign the trait a map unit position on a chromosome. A LOD score of less than 3.0, with a recombination frequency of approximately 50%, indicates a high probability that the trait and probe are unlinked. These techniques have been used to pinpoint the loci for a number of different single genes that cause disease in humans. In addition to cystic fibrosis, loci for Huntington's disease, Duchenne muscular dystrophy, Alzheimer's disease, and Parkinson's disease, among many others, have been identified. These traits are a few of the 22 traits that you can study in PedigreeLab. And pinpointing the location for a rare trait is the ultimate goal that you will strive for using PedigreeLab!

With PedigreeLab you will choose a gene from a list of mutant genes that cause actual genetic disorders in humans. Most of these disorders are relatively rare. To determine the location of the gene for the trait that you are following in PedigreeLab, you will first study sample pedigrees to develop a hypothesis on the mode of inheritance for the trait. You will then search large family databases of RFLP data that were generated with DNA probes to different markers on human chromosomes. One of your challenges will be to find a probe for a marker that is linked to the gene for the trait that you are studying. Once you have done this, you will analyze RFLP and pedigree data from families that you will choose as ideal families for generating recombination data. You will be looking for recombination between the locus for the marker and the trait that you are studying. This will be a particularly challenging aspect of PedigreeLab because you must consider parents with the proper arrangement of the marker and trait on a set of homologous chromosomes to be able to detect recombination in the offspring.

With PedigreeLab you will count recombinants and nonrecombinants based on RFLP analysis and the genotypes that you have assigned to individuals in a pedigree. Recombination frequency and LOD values are tallied for you. If you see a LOD score of >3.0, then it is likely that the marker you chose and the trait are linked. You can then use the recombination frequency data to assign a position for the trait on the same chromosome as the marker. If, however, your LOD score is <3.0, and recombination frequency is approximately 50%, then the probability is high that the trait and marker are unlinked and you'll need to start your gene hunting again with a different probe!

References
1. Cummings, M. *Human Heredity*, 3rd ed. St. Paul, Minn: West, 1994.

2. Lander, E., and Botstein, D. "Mapping Mendelian factors underlying quantitative traits using RFLP linkage maps." *Genetics* 121 (1989).

3. Lewin, B. *Genes VI.* New York: Oxford University Press, 1997.

Introduction

In FlyLab you studied modes of inheritance and chromosome maps in *Drosophila melanogaster*. In this laboratory you will use pedigree analysis to simulate the inheritance of genes for human genetic disorders and RFLP analysis to study recombination in humans. Combining RFLP analysis with pedigree analysis will enable you to learn about, and understand, how the location of a gene can be assigned to a chromosome. You will also pinpoint the location of a gene on a chromosome by using recombination data to generate a genetic map of that chromosome.

Objectives

The purpose of this laboratory is to:
- Demonstrate how pedigree analysis can be used to determine the mode of inheritance for a genetic disorder across several generations of humans.
- Demonstrate how recombination data from genetic crosses can be interpreted and applied to discover the location of a gene on a chromosome and to develop chromosome maps.
- Simulate the use of restriction fragment length polymorphism (RFLP) analysis to develop genetic maps of a chromosome.

Before You Begin: Prerequisites

Before beginning PedigreeLab you should be familiar with the following concepts:
- Chromosome structure and the stages of gamete formation by meiosis (see Campbell, N. A., Reece, Reece, J. B., and Mitchell, L. G., *Biology* 5/e, chapters 13–15).
- Using FlyLab to study basic principles of Mendelian genetics.
- Inheritance of normal traits and genetic diseases by autosomal recessive, autosomal dominant, and sex-linked inheritance in humans (chapters 14 and 15).
- Use of genetic crosses to develop chromosome maps according to linkage groups (chapters 14 and 15).

Assignments

Twenty-two different mutations of human genes that produce actual genetic diseases are included in PedigreeLab. Some of these mutations produce diseases that you have probably heard about and already studied; in other cases, rarer genetic mutations and diseases are presented. You will use PedigreeLab to study a number of different pedigrees to develop a hypothesis to explain the mode of inheritance for the trait that you selected. Once you have done this, your goal is to study recombination between the mutant gene and markers that you will select, using RFLP analysis to help you determine the chromosome containing the mutant gene, and the location of the mutated gene relative to genetic markers on that chromosome.

Getting to Know PedigreeLab: Inheritance of Amyotrophic Lateral Sclerosis (ALS)

The first screen that appears in PedigreeLab is the Information view, beginning with Alzheimer's disease. Click on the popup menu for the mutation list. A list of abbreviations for gene mutations of various human genetic disorders appears. Scroll down this list and view the different mutations and diseases presented. Note that for each mutation you are provided with a short narrative of background information describing the history, phenotypes, and other symptoms created by the mutation. You are not provided with any clues, however, about the location of the gene for this condition. It is your job to hunt for the chromosomal location of each mutation that you choose! Choose the ALS abbreviation for the mutated gene that results in amyotrophic lateral sclerosis, also commonly known as Lou Gehrig's disease, named after the famous New York Yankees baseball player who died from the condition.

Read the information panel to learn more about ALS. You may have studied ALS before, and you may also know that another former Yankee, Jim "Catfish" Hunter, was recently diagnosed with ALS. This first assignment is designed to introduce you to the process and intricacies of using PedigreeLab by studying the inheritance of ALS and mapping the locus of the ALS gene to its correct chromosome. The other assignments in this laboratory are based on your ability to use your understanding of genetics, and the basic functions of PedigreeLab that you will learn in this first assignment, to map the mutations in PedigreeLab.

1. <u>Determining the Mode of Inheritance for a Gene by Studying Full Pedigrees</u>
   To determine the pattern of inheritance for a particular mutation, click on the Full Pedigree tab at the top of the screen. The pedigree that is displayed on this page is one of 100 sample pedigrees for ALS available in PedigreeLab. To choose a different pedigree, click on one of the up or down arrows in the upper right corner of the screen. The purpose of this Full Pedigree function is to provide you will a large number of different pedigrees that you can evaluate to develop a hypothesis for whether a particular mutation is inherited by autosomal recessive, autosomal dominant, or sex-linked inheritance.

   a. When studying these pedigrees, remember that circles represent females and squares represent males. Parents are connected by a red horizontal line. Black vertical lines from the parents indicate the offspring produced by these parents. Sibs are connected to each other by a horizontal black-lined bracket shown above the symbols. The symbol for an individual that expresses the mutant phenotypes is shaded blue. First-generation parents are shown at the top of the pedigree, followed by the second generation, third generation, and so on.

b. You may already know that ALS is inherited as an autosomal dominant condition. Heterozygotes develop the disease while homozygous recessive individuals are phenotypically normal. Homozygous dominant individuals rarely live to adulthood because of the lethality of this mutation. Study several of the full pedigrees until you are comfortable with the inheritance of ALS as an autosomal dominant trait.

c. <u>Assigning Genotypes to the Individuals in a Pedigree</u>
Select a simple pedigree with a first-generation parent that has ALS and a first-generation parent who is phenotypically normal. You can assign genotypes to each of the members of a pedigree as follows:

   (1) To assign a genotype to each parent, double-click on the circle for the first-generation mother. A new box should open prompting you to type in the mother's genotype. Type in "Aa" or "aa" depending on whether the mother is a heterozygote with ALS or a phenotypically normal homozygote, then click OK to close the genotype box. Double-click on the square for the first-generation father and type in the appropriate genotype for the father, then click OK to close the genotype box. The genotype for each parent should appear below each symbol.

   (2) Repeat this process to assign genotypes for the second- and third-generation individuals in the pedigree. Once you have filled in the genotypes for a pedigree, the labeled pedigree can be exported as a GIF file by clicking on the Export Pedigree button at the lower right corner of the screen. A new browser window with your labeled pedigree will appear. You can now print your pedigree from this window. Analyze at least two other pedigrees to confirm your genotype assignments.

2. <u>Mapping the Locus for a Mutant Gene by Studying Large Family Pedigrees, RFLP Analysis, and Recombination</u>
Identifying the chromosome that contains the mutation that you are studying and mapping the locus for this mutation using PedigreeLab involves your ability to apply what you learned about the mode of inheritance for the mutation, and your ability interpret the results of RFLP and recombination data. This process is part of the challenge and fun of PedigreeLab.

Using the Large Family function is the key to the gene-mapping functions of PedigreeLab. This function simulates databases that present family pedigrees for individuals with different genetic disorders that have been subjected to RFLP analysis using one of 17 different probes for markers on human chromosomes. The pedigrees are arranged according to genotypes for the RFLPs; phenotypes are also shown with these pedigrees. The probes in PedigreeLab do not correspond to actual marker loci in human chromosomes.

You will pick an appropriate pedigree that will allow you to look for recombination between the marker sequence of DNA recognized by the probe and the mutation. Based on recombination frequencies that you generate, and LOD scores, you will then decide whether the probe and mutation are linked to the same chromosome. If they are linked, you will use the Chromosome View function to pinpoint the locus for your trait based on recombination frequencies. All of the traits presented in PedigreeLab will map to their actual chromosomes as found in real life. No shortcuts or hints are given to determine linkage groups; hence, you must be a gene hunter using trial and error and analysis of your data to eliminate loci for unlinked probes and find linked probes that will help you pinpoint the locus for the mutant gene!

a. Begin by clicking on the Large Family tab at the top of the screen. Before you view the family pedigrees, answer the three questions on the expression of the mutant phenotype in a pedigree. You have the option of selecting whether the mutant trait is expressed in grandparents, parents, and offspring. The purpose of these three questions is to better define and narrow your search of the Large Family databases to select for those pedigrees that will be most helpful to study recombination.

   (1) Remember that for most traits you will want to define the search to look for pedigrees that have one parent who is a double heterozygote for the trait and probe (marker) and one parent who is a double homozygote for the trait and probe.

   (2) For this ALS assignment, answer yes to each question. For the other assignments, think carefully about the mode of inheritance of the trait that you are working on before answering yes or no to each of the three questions. Before searching the database, note that the default probe selected should be the "asr" probe. Click on the Search Database button. Your search of the database will reveal different family pedigrees that are assigned genotypes resulting from RFLP analysis with the asr probe.

b. <u>RFLP Analysis</u>
   The first screen that appears after searching the Large Family database displays a pedigree in which individuals are labeled according to their phenotype expressed and assigned a genotype according to the results of RFLP analysis with the asr probe.

   (1) To interpret each phenotype, note that the symbols for individuals expressing the mutant phenotype are shaded blue. You can assign a genotype for the trait to each individual in the pedigree by double clicking on each symbol as you did with the Full Pedigrees and entering the genotypes for each individual in the pedigree.

(2) To interpret the results of the RFLP analysis, genotypes are labeled in each symbol of the pedigree according to the presence (+) or absence (–) of a restriction site in the fragments of DNA detected by the probe. Individuals who are homozygous for the presence of a restriction site are labeled as "+/+." Heterozygous individuals are labeled as "+/–," indicated the absence of a restriction site on one homologue and the presence of a restriction site on the other homologue. Homozygous individuals lacking a restriction site on both homologues of a pair are labeled as "–/–."

(3) Notice that an autoradiogram showing the RFLP pattern that would appear for different individuals depending on their RFLP genotype is shown on the right side of the screen. Reading the autoradiogram from left to right, the first lane, labeled "Ref," consists of a series of DNA fragments of known size in kilobases (kb) that are used as size standards, or references for determining the size (in base pairs) of the RFLP fragments. Clicking on the symbol for any individual in the pedigree will show their RFLP pattern in the "selection" lane of the autoradiogram. Compare the size of each DNA fragment with the size standards; note that large DNA fragments are located at the top of the autoradiogram while smaller fragments migrate relatively faster toward the bottom of the autoradiogram. You may want to refer back to Figure 1 in the background section of this lab to help you interpret the RFLP autoradiograms.

c. <u>Choosing a Large Family Pedigree to Study Recombination Between the Trait and Marker</u>
Once you are comfortable with how a Full Pedigree is shown, use the arrows in the upper right corner of the screen to find a pedigree that is appropriate for studying recombination. Begin by looking for a pedigree with one parent who is a double heterozygote (A/a, +/–) for the trait and probe, and a parent who is a double homozygote for the trait and probe (a/a, +/+). Look carefully at the phenotype and RFLP genotype of the grandparents before assigning genotypes to the parents. Once you have found an appropriate pedigree, you may find it helpful to label all of the individuals in the pedigree before attempting to count recombinant and nonrecombinant offspring. *Hint*: If you are having difficulty finding the correct pedigree, use the Genetic Calculator function described below.

(1) The **Genetic Calculator** function can be very useful for helping you determine which pedigrees will be best suited for studying recombination between the trait and marker. This function allows you to specify genotypes in the parents of a full pedigree that show the trait and marker linked to a pair of homologous chromosomes or unlinked on different chromosomes.

To use this function, click on the Genetic Calculator tab at the top of the screen, then click on the popup arrow next to the box labeled "Trait & probe autosomal, linked" at the top of this window. This popup menu allows you to design genotypes for different individuals depending on whether you think that the trait you are trying to map is an autosomal or sex-linked trait, and whether or not this trait is linked or unlinked to the probe for a given marker. Because you know from studying the full pedigrees that ALS is inherited as an autosomal dominant condition, select "Trait & probe autosomal, linked" to design genotypes that would show the ALS trait and asr marker as linked on an autosome.

Notice that on the far left of this box there are four popup menu boxes that you can use to select the trait and marker combination for each homologue of a pair in a female parent. Click on one of the upper two boxes: These allow you to choose the genotype for the trait. Note that the dominant allele for ALS is shown in capital letters and the recessive allele for ALS is shown in lowercase letters. Click on one of the lower two boxes: These allow you to choose the genotype for the probe (marker). Note that the genotype for a homologue with a restriction site is shown as "asr+," and the genotype for a homologue lacking a restriction site is shown as "asr–."

It is important that you determine the correct linkage combination for the parent who is a double heterozygote (A/a, +/–) for the trait and restriction site. Do this for the female parent. For the first homologue, select a linkage combination with "ALS" and "asr+" on the first homologue. Select the opposite homologue with the linkage configuration of "als" and "asr–."

Repeat this process for a male parent who is a double homozygote for the trait and restriction site (a/a, +/+). For this parent, the linkage combination is the same on both homologues.

Once you have selected the genotypes for each parent, click on the Calculate Punnett Squares button. You should now see Punnett squares that show nonrecombinant and recombinant genotypes that would result from a cross between the parents with the trait and restriction site arrangements that you just set up. You are now ready to return to the Large Family pedigree function to look for a pedigree with parents that show the genotypes for the trait and marker combination that you set up with the Genetic Calculator. Once you have done this, use the results of the Punnett squares to help you look for the genotypes that you will count as recombinants and that you will count as nonrecombinants. *Hint*: To help you as you search the Large

Family pedigrees, you may want to export the results of your Punnett squares from the Genetic Calculator and print a copy of these results to have in front of you as you look at the large family pedigrees to count recombinants and nonrecombinants.

(2)   <u>Counting Recombinants and Nonrecombinants</u>
Once you have located a Large Family pedigree with the double heterozygous and double homozygous parents that you determined with the Genetic Calculator function, begin to tally recombinant and nonrecombinant offspring by clicking on the arrow buttons at the bottom of the screen.

After counting offspring for one pedigree, choose another useful pedigree and look for recombination. Note that as you tally recombinants and nonrecombinants, a running tally for the total number of offspring that you have counted is shown at the bottom of the screen along with recombination frequency and LOD score. You will need to count offspring from several pedigrees to accumulate data that will help you determine if the trait and probe are linked. Remember that you are looking for a LOD score of 3 or greater, and a recombination frequency of less than 50% as good evidence that the trait and probe are linked. If you are seeing evidence for linkage, then you are ready to map the trait to a chromosome.

Conversely, a recombination frequency of 50% indicates that the probe and trait are probably unlinked. This can usually be determined after analyzing just a few pedigrees. If this is the case, choose another probe and try again! You should find that recombination data using the asr probe indicates that the asr marker is not linked to the ALS gene. Click on the Chromosome View and note that the asr probe is located on chromosome 1; therefore, you can probably rule out this chromosome as the location of the ALS gene. You may also want to rule out using the other probes on this chromosome in your next search. You may find that you will have to search for recombination with several probes before you find one that is linked to your trait.

To look for linkage with another probe, return to the Large Family function. Click on the New Search button. This time, search the database for pedigrees using the "juva" probe. Once you have done this, count recombinant and nonrecombinant offspring again. This time you should see evidence for linkage of the juva probe and ALS gene.   Follow the next procedure to map the ALS gene to the correct chromosome.

d. <u>Mapping the Mutation to a Chromosome</u>
Now that you have evidence for linkage, click on the Chromosome View tab at the top of the screen. Genetic maps appear that show loci for markers on five different chromosomes. Notice that the juva marker is located on chromosome 21. To position the ALS gene, click and drag the ALS arrow on the upper right side of this window. Drag the ALS arrow to chromosome 21 and position the arrow relative to the juva marker according to the recombination frequency value that you generated with the Large Family pedigrees. For example, if you generated a recombination frequency of 15%, you should move the ALS arrow to a position that is 15 map units (centimorgans) away from the juva locus. You will not know, however, on which side of the juva locus to position the ALS arrow until you study linkage between the ALS locus and the other markers on chromosome 21.

Notice the other markers on chromosome 21, "ddy" and "utc3." Return to the Large Family function and repeat a database search using probes for each of these markers to better pinpoint the location of the ALS gene. The more offspring you tally, the more reliable will be your estimates of the recombination frequencies. Counting 50–100 offspring for each probe will give you good values. You can also refine your estimate by averaging the map locations for the different probes. You should find that the ALS gene maps to a position just above the juva marker. See your instructor for the exact location of the ALS gene.

e. <u>Printing Your Chromosome Map and Recombination Data</u>
Print a copy of the chromosome map that you developed by clicking on the Export Graphic button in the lower right corner of the screen. A new GIF image window will open which you can now use to print your map.

To print your recombination data, return to the Large Family view. You can export a single pedigree that you have labeled by clicking on the Export Pedigree button and printing from the GIF image window that appears. You can also export all of the data (total number of recombinants, nonrecombinants, recombination frequency, and LOD scores) for each probe that you used by clicking on the New Search button, then clicking on the Export Whole List button. With this function you will open a window with your data that functions as a lab notebook. You can add any comments that you'd like to your data and print a copy of this data as evidence of your gene-hunting progress.

On Your Own: Mapping Other Mutations

Now that you have experienced how PedigreeLab functions to map the location of the
ALS gene, you are on your own to be a gene hunter for some of the other mutations in
PedigreeLab!

1. Follow the guidelines in the ALS assignment to help you identify the location
   and approximate locus for each of the following mutations:

   - Huntington's disease (HD)
   - Duchenne muscular dystrophy (DMD)
   - Werner's syndrome (WRN)
   - Cystic fibrosis (CF)
   - Two additional mutations of your choice.

2. For each mutation, perform the following:

   a. Study several Full Pedigrees, then develop a hypothesis to explain the
      mode of inheritance for the mutation.

   b. Use the Large Family pedigree function to identify pedigrees with
      parents that will produce offspring that you can use to look for
      recombination. Think carefully about the mode of inheritance for the
      trait, then decide whether to answer yes or no to the three questions that
      ask you to define whether the trait is expressed in grandparents,
      parents, and offspring before searching the Large Family database.

      Use the Genetic Calculator to help you select the correct Large Family
      pedigrees and decide what genotypes to look for when counting
      recombinant and nonrecombinant offspring. Once you have selected
      the correct pedigree, count the number of recombinant and
      nonrecombinant offspring. Do this until the recombination and LOD
      data indicate whether the trait and probe are linked or unlinked. If the
      trait and probe are linked, use the Chromosome View function to assign
      a position for the mutant gene on the correct chromosome. If the trait
      and probe are unlinked, then pick another probe to test for
      recombination and repeat the process until you find a probe that is
      linked to the trait.

   c. Print your recombination data for any probes that you used to search
      for linkage.

   d. Once you have found the chromosomal location for the mutation, be
      sure to test all of the other probes on that chromosome that are linked to
      the mutation so you can better refine and pinpoint the location of the
      trait relative to the markers on the chromosome. Your goal is to
      develop as accurate a map as you can. Print your chromosome map
      when you are done.

e. Show your recombination data and chromosome maps to your instructor to see how accurate your mapping experiments were.

Group Exercises

The Human Genome Project is a worldwide initiative to develop genetic maps for the 22 autosomes and two sex chromosomes in humans. One long-term goal of the Genome Project is to identify and determine the nucleotide sequence for an estimated 100,000 genes that are thought to be present in the approximately three billion nucleotides of DNA that comprise the human genome. The Genome Project began in 1990 and was targeted as a 15-year project with an estimated cost of $3 billion. Although this project originated in the United States, labs around the world are contributing to the Genome Project. Many new technologies for identifying and sequencing genes have been developed as a result of the project. Because of this, the progress of the Genome Project is actually slightly ahead of schedule and just under budget. One technique being used by Genome Project scientists involves using RFLP analysis and other similar techniques to develop extensive genetic maps of chromosomes in humans and other animal and plant species such as the nematode *Caenorhabditis elegans*; *Drosophila melanogaster*; the common house mouse, *Mus Musculus*, and the flowering plant *Arabidopsis thaliana*. Some labs involved in the Genome Project are working on mapping assigned segments of certain chromosomes. Other groups are involved in checking the accuracy of maps completed to date.

The following exercises are designed to have you investigate the accuracy of genetic mapping as simulated in PedigreeLab, and to enable you to map an entire chromosome. Work together in a group of four students to complete the following assignments.

1. Select two mutant genes that you have not mapped already. Divide your group into pairs, and have each pair of students work together to identify the chromosome and approximate locus for each of the two mutations. Once each pair has mapped the two genes, compare your results with each other. Did each pair in your group identify the correct chromosome(s)? Why or why not? If you cannot agree on the correct chromosome, go back and review your work until everyone in your group can agree on a correct chromosome for each mutation.

    Once you have agreed on a chromosome location, compare the accuracy of the locus that you identified for each mutation. How accurate were the maps when you compared them? How many map units did the two maps differ by? Explain possible reasons for any differences in loci assigned on each chromosome mapped.

2. You have been selected as the leading scientist directing a group of scientists responsible for mapping the X chromosome. Pick three other students who can work as your partners to narrow down your hunt to map the X chromosome.

How would you begin? Should you assign each student a mutant gene that you have not yet mapped and have each person study the inheritance for that gene? Or should you work as a group and look for inheritance patterns to first narrow your search for genes that appear to be X-linked? Decide on an approach to use, then work together to develop a map of the X chromosome that will show all of the X-linked mutations present in PedigreeLab. Because PedigreeLab will show different regions of the X chromosome depending on the probes you are using, PedigreeLab will not show one complete full-length map of the X chromosome. Your group will have to piece together the different maps of the X chromosome shown in PedigreeLab to draw one complete map.

Once you have done this, compare your map with that of another group of students in your class that has worked on the X chromosome. How accurate were the two sets of maps? Were there any differences in genes mapped to the X chromosome when the maps were compared? If there were differences, work together with the other group to resolve these differences until you can agree on an accurate and complete map of the X chromosome. Print your map and show it to your instructor for comparison with the correct map.

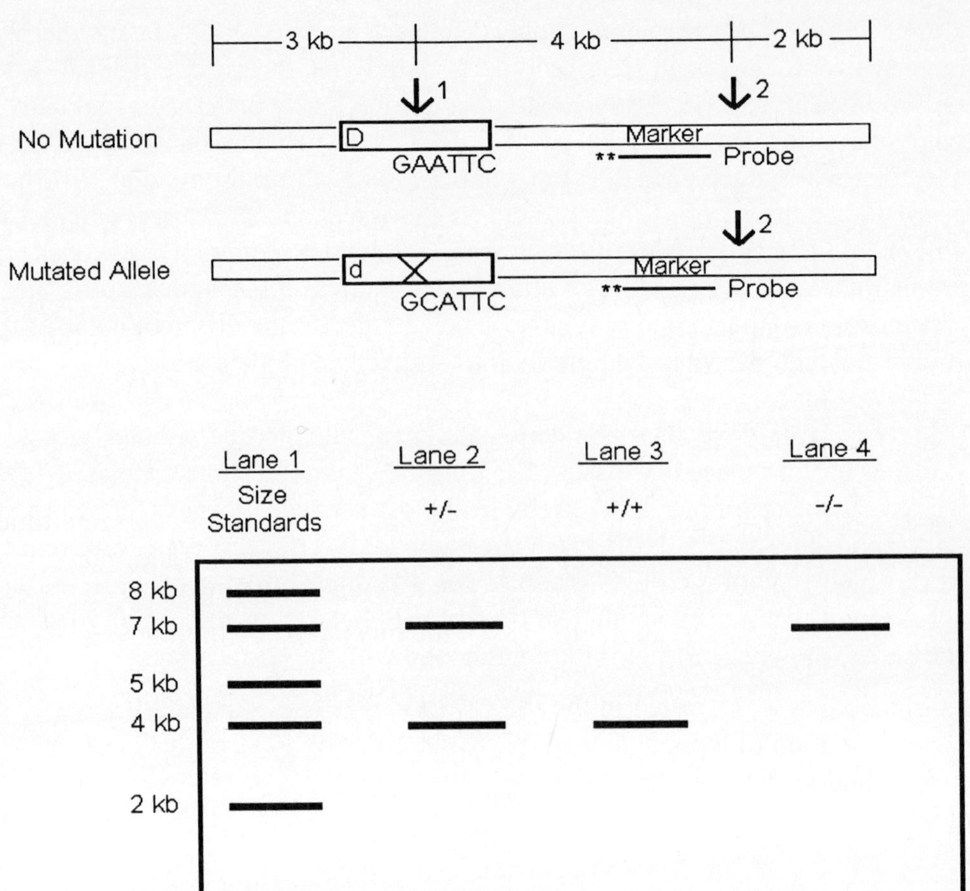

Figure 1: Diagrammatic Representation of Restriction Fragment Length Polymorphisms (RFLPs) for a Recessive Trait. Shown is an example of two homologues from an individual who is heterozygous (Dd) for a recessive gene that is linked to a genetic marker. Arrows indicate the restriction enzyme recognition sequence that would be cut by restriction digestion of each chromosome with the enzyme *Eco*RI. Note that two sites (1, 2) are present on the homologue with the dominant allele (D), while only one site (2) is present on the homologue with the recessive allele (d). The absence of restriction enzyme cutting site number 1 in the recessive allele due to a single nucleotide mutation is indicated with an "X." The location of a radioactive probe that is complementary to the marker sequence is also shown. The pattern of restriction fragment length polymorphisms that would be generated for these homologues is visualized after gel electrophoresis of the digested DNA fragments, transfer of the DNA fragments to a filter, and hybridization of the fragments to the radioactive probe, followed by exposing the filter to X-ray film to produce an autoradiogram. In the autoradiogram, lane 1 shows the migration of DNA size standards of known length that are used to determine the nucleotide length of each RFLP. Lane 2, shows RFLPs for a heterozygote (+/–) where the genotype is designated according to the presence (+) or absence (–) of a restriction site; cutting with *Eco*RI yields two fragments, 7 kb and 4 kb in size. For comparison, lanes 3 and 4 show RFLPs for individuals who are homozygous for the restriction site, +/+ or –/–.

DemographyLab

Background

Ecology includes a broad and fascinatingly complex range of topics related to the interrelationships between living organisms and their nonliving environments. Ecologists classify these interactions into a hierarchy based on the complexity of interrelationships that occur between living (**biotic**) components, and nonliving (**abiotic**) components (such as soil, water, weather, temperature, pollutants, light, and nutrients). At the simplest level of this hierarchy, organismal ecology focuses on the interactions between individual organisms and the environment. Individual organisms can be grouped into a **population**, which is defined as a group of individuals of the same species that occupy and are adapted to a particular geographic area at the same time. Populations of different species that inhabit a particular area at the same time are grouped together into **communities**. All of the organisms and nonliving components within a given area are classified as an **ecosystem**. At the highest level in this hierarchy, all ecosystems on Earth are part of the **biosphere**.

Population biologists are interested in the many factors that influence the size, health, growth, and distribution of a population. The scientific study of the factors that affect population size and growth over time is called **demography**. Two important factors that affect population growth are population **density** and the spacing or **dispersion** of individuals within the geological boundaries in which a population lives.

Ecologists have established a variety of methods for estimating the density of a population and for studying density-dependent and density-independent factors. Depending on the abundance and wariness of a species, sampling techniques can include signs of an organism such as nests, droppings, tracks, and reduction of specific forage.

Ecologists who study dispersion are particularly interested in the patterns of spacing that many species exhibit within a particular geographical range. Interactions of individuals within a population and the influence of abiotic factors will affect the spacing patterns of individuals within their geographical range. These patterns can involve grouped (clumped), evenly spaced (uniform), and unpredictable (random) dispersions of organisms.

Both density and dispersion of a population are directly influenced by three primary factors: (1) reproduction or birth rate (fertility), (2) death rate (mortality), and (3) migration. In migration, the addition of an organism to a population is known as immigration while the loss of an organism from a population is called emigration. Population growth remains the same (zero population growth; ZPG) if the number of individuals that join a population, either by immigration or by birth, is equal to the number of individuals that leave a population by either emigration or death. If the number of births and immigrating organisms exceeds the number of deaths and emigrating organisms, then population growth occurs. The reverse is true when a population is on the decline. In addition, both the birth rate and death rate of a population rely on the age of its members (**age structure**). For example, a population that has a higher percentage of older individuals that may be past their reproductive prime compared to younger individuals is likely to show a higher degree of mortality compared to birth rate.

Given the aforementioned factors and additional factors such as generation time and sex ratios, biologists, demographers, and sociologists can use statistical data and computer programs to predict future changes in the size of human populations. When studying populations of other organisms, factors such as competition and predator-prey relationships must also be considered.

Human population growth has wide-reaching impacts on human health and the infrastructure, resources, and carrying capacity of a given geographic region. The importance of understanding human population growth is exemplified by the following quote from the opening chapter of *The Population Explosion*, by Paul and Anne Ehrlich: "The population of the United States is increasing much more slowly than the world average, but it has more than doubled in only six decades; from 120 million in 1928 to 240 million in 1990. Such a huge population expansion within two or three generations can by itself account for a great many changes in the social and economic institutions of a society. It is also very frightening to those of us who spend our lives trying to keep track of the implications of the population explosion."

DemographyLab is designed to simulate how different demographic factors may influence human population size and growth, although most of the factors that you will consider in this lab have similar effects on the population growth of other animals and plants.

References

1. Kates, R. W. "Sustaining Life on the Earth." *Scientific American*, October 1994.

2. Raloff, J. "The Human Numbers Crunch." *Science News*, June 1996.

3. Ehrlich, P. R., and Ehrlich, A. H. *The Population Explosion*. New York: Simon & Schuster, 1990.

Introduction

DemographyLab will help you learn about the current demographics for human populations of seven countries. You can use this lab to investigate how population size, population structure, mortality rate, and fertility rate influence the size and growth of human populations over time. For simplicity, the effects of immigration and emigration are not included in DemographyLab. However, because the demographic data presented in this lab have been determined from actual current demographic statistics for each country, the population changes you will examine may simulate future demographic realities for these seven countries.

Objectives

The purpose of this laboratory is to:

- Demonstrate fundamental primary demographic factors that influence human population size and growth.
- Simulate the effects of demographic factors on human population growth.
- Use demographic models to predict future changes in human population growth.

Before You Begin: Prerequisites

Before beginning DemographyLab you should be familiar with the following concepts:

- The four major levels of study in ecology based on the complexity of interactions between organisms and their environment (see Campbell, N. A., Reece, J. B., and Mitchell, L. G., *Biology* 5/e, chapter 50).
- Important determinants of organism distribution within the biosphere (chapters 50 and 52).
- The definition of demography and examples of important factors that affect human population size and growth (chapter 52).

Assignments

Demographic Differences Among Nations

Countries differ with respect to population numbers, age structure, and fertility and mortality rates. These differences are caused by many factors, such as geographic size and location, level of economic development, government policies, and religious practices. The following exercises are designed to help you understand the influence of some of these factors on the size and growth of human populations.

1. Select the Population Structure view on the input screen of Demography Lab. Using the Country popup menu, examine the estimated 1998 population structure of each nation. Can you classify the population structures into two general patterns? Consider what you know about each of these countries. What do you think is the biggest factor distinguishing these two groups of nations?

2. Select the Fertility Rate view on the input screen of DemographyLab. Using the Country popup menu, examine the estimated 1998 fertility rates of each nation. (Try changing the scale to magnify these differences.) Do you see any trends for the fertility rates compared to the population structures? What are they?

3. Select the Mortality Rate view on the input screen of DemographyLab. Using the Country popup menu, examine the estimated 1998 mortality rates of each nation.

a. Do you see any general similarities and differences between nations? Provide possible reasons for these similarities and differences. *Hint*: To help you determine differences in mortality rates, click on the Run button and choose the Vital Rates view. Export the data from this table to your notebook by clicking on the Export Data button. Click OK to export data to your notebook as text.

b. Do you see any trends for the mortality rates compared to the age structures? What are they? Which age groups show the biggest differences in mortality rate among nations?

c. Do you see any differences in the mortality rates of males compared with females for any of the countries? What are they? What might account for some of these differences?

4. Most scientists believe slower population growth is beneficial to the sustainability and quality of human life on Earth. In the mid-1980s, the World Health Organization (WHO) Task Force on Contraceptive Vaccines specifically designated a large sum of research funds for scientists to design novel contraceptive methods that could be used to slow population growth. The WHO identified certain developing nations of Africa, South America, and Asia, including India, as populations that could benefit from new contraceptive approaches (such as contraceptive vaccinations) that could easily be administered on a large scale.

Select the Fertility Rate view on the input screen of Demography Lab. Examine the fertility rates for Nigeria and India.

a. Develop a list of differences and similarities that appear for the fertility rates of each country. Why might each country show these differences and similarities?

b. Can you think of common societal issues shared by India and Nigeria that might account for the similarities?

c. Provide possible explanations for why it may be difficult to establish contraceptive techniques in these countries that would have a significant impact on population growth.

Historical Effects on Demographic Changes

For a variety of reasons, populations sometimes undergo periods when there are large changes in reproduction or mortality. For example, desert plant populations may produce a large cohort of seedlings following heavy rains, or a fish population may experience a peak in reproduction due to a La Niña weather year. Human populations exhibit similar demographic changes. For example, the history of human civilization includes many well-documented events, such as plague (Black Death of medieval

Europe) and famine (Irish Famine of 1845–1847), that produced pronounced demographic changes in human populations. The following exercises illustrate some examples.

1. In the United States, a peak in reproduction occurred after World War II from the late 1940s through the early 1960s; this produced the "baby boom" generation. Select the Population Structure view on the input screen of DemographyLab. Using the Country popup menu, examine the estimated 1998 population structure of the USA. Can you find the baby boomers?

   Click the Run button and choose the Population Structure view. (You may wish to use the Scale button to make the differences in age structure more obvious.) Use the arrow buttons at the top of the view to advance the population structure forward in time. Follow the baby boom generation as they age. Compare the population in 1998 with the projected population for 2028. What do you think the social consequences will be for these changes? Provide at least three examples of social consequences that will result from these changes.

2. In 1980, China's government adopted a policy advocating one child per couple. Select the Population Structure view on the input screen of DemographyLab. Using the country popup menu, examine the estimated 1998 population structure of China. Can you see evidence of China's population policy?

   Click the Run button and choose the Population Structure view. (*Hint*: You may wish to use the Scale button to make the differences in age structure more obvious.) Use the arrow buttons at the top of the view to advance the population structure forward in time by 5-year increments. What changes will occur in China's population structure as a result of its policy?

3. At the same time the baby boom was occurring in the United States, there was a "baby bust" in post-World War II Japan, when fertility rates decreased. Select the Population Structure view on the input screen of DemographyLab. Using the Country popup menu, examine the estimated 1998 population structure of Japan. Can you see evidence of the postwar baby bust? What has happened to fertility rates in Japan in recent years?

   Click the Run button and choose the Population Structure view. (You may wish to use the Scale button to make the differences in age structure more obvious.) Use the arrow buttons at the top of the view to project the population structure forward in time. If you were a demographer, what advice or concerns would you have for Japan's government in planning for the future?

Stable Age Structure

If vital rates remain constant, what will happen to population structure over time? This question is investigated in the following exercises. Keep in mind that although we are working with a human population model, the same principles will apply for any population of plants or animals

1. Set the number of years to 300 and simulate the population growth of each of the seven nations by using the Population Structure view as described in the previous assignment. Examine the pyramid plots as they change every five years. What happens to each population over the long term? How do the plots differ for each nation? Which nations show the biggest changes over time? Speculate about why these changes occur.

2. If a population's age structure stops changing from year to year, demographers say that it has a stable age structure. Two typical patterns of stable age structures are pyramids and inverted pyramids. Compare the stable age structures and intrinsic growth rates of the seven countries. Do you see a pattern? Develop a hypothesis about the relationship between growth rate and the shape of the stable age structure. Test your hypothesis by selecting a nation and altering its fertility rates. Can you create the pyramid and inverted pyramid patterns?

3. Suppose the human immunodeficiency virus (HIV), the virus that causes acquired immunodeficiency syndrome (AIDS), were to mutate into an airborne virus that could easily be transmitted by casual contact with infected individuals. Choose any country, decrease its 1998 population size based on mortality due to this new strain of HIV, and simulate the population for 300 years. What effect, if any, does this have on the long-term age structure? Examine the pyramid plots as they change every five years.

    Repeat the procedure for alteration in the population structure by decreasing the population of sexually active males and females in the 20-year-old categories. Examine the pyramid plots as they change every five years.

    Repeat the procedure for fertility rates, and for female and male mortality rates. Formulate a hypothesis for how changes in each of these factors affect the long-term population structure. Design and carry out simulations to test your hypothesis.

Zero Population Growth

Zero population growth (ZPG) is of special interest to demographers and ecologists because population numbers will remain constant over time. The net reproductive rate (NRR) is a good indication of how close a population is to ZPG, because it represents the average number of daughters that a woman is expected to produce in her lifetime. ZPG is attained when NRR equals one; every woman is exactly replacing herself. (It is not necessary to consider males, from a demographic point of view, since only women bear children.) If $NRR > 1$, a population will increase and if $NRR < 1$, a population will decrease. A value of $NRR = 1$ corresponds to an intrinsic growth rate of zero. ZPG is investigated in the following exercises.

1. Simulate each country's population for 300 years and look at the time series plot of the population size. Form a hypothesis about which countries are

closest to ZPG (that is, NRR = 1). Compare the NRRs of the seven countries.
How well did you do with your hypothesis?

2. Choose a country with a high population growth rate (for example, Nigeria)
    and modify its fertility rates until you get as close as possible to ZPG. What
    types of changes to fertility rates were necessary to attain ZPG? Reset fertility
    rates back to their default values and change female mortality rates until you
    get as close as possible to ZPG. What types of changes to female mortality
    rates were necessary to attain ZPG? Do the same changes to male mortality
    rates achieve ZPG? Why or why not? Given the types of changes necessary to
    achieve ZPG, what realistic options are open to a nation to reach ZPG?

Demographic Momentum

Even though a country may have fertility and mortality rates that give a zero or
negative growth rate, population size may increase for many years before leveling off
or decreasing. Demographers call this demographic momentum. The following
exercises illustrate this phenomenon.

1. Simulate population growth in China for 100 years. How long does it take for
    the population size of China to begin decreasing? Does any other country
    exhibit a similar phenomenon?

2. Choosing a country with a negative growth rate, try modifying its 1998
    population structure so it will exhibit demographic momentum.

3. Propose a hypothesis about what causes demographic momentum. Test your
    hypothesis by altering the default values of any country and simulating it in
    DemographyLab. Design an experiment to illustrate negative demographic
    momentum (a decrease in population size followed by a long-term increase).

Dependency Ratios

From a social and economic point of view, human demographers are concerned about
the proportion (ratio) of workers in a population to those who are economically
dependent on these workers, either directly or through taxes such as Social Security.
the following formula, called the dependency ratio, is used as a summary statistic.

Dep. Ratio = 100 [(no. children under 15) + (no. elderly over 65)] / [no. persons 15
– 64]

We look at the dependency ratio in the following exercises.

1. Referring to the Tabular Data view of DemographyLab, compute the 1998
    dependency ratio for each of the seven countries. Compare and contrast each
    nation. Which component, children under 15 or elderly over 65, is most
    important for determining the dependency ratio for different nations?

2. Compute the dependency ratio for the USA for the years 1998, 2003, 2008, 2038. What happens to the dependency ratio as the USA enters the next century? Which component, children under 15 or elderly over 65, is most important for determining the changes in the dependency ratio? What are the social or economic implications of these changes?

3. What do you think will happen to the dependency ratio for a country over the long term if vital rates remain constant? Test your hypothesis using DemographyLab.

Sex Ratios and the Marriage Squeeze

The sex ratio for a population or age group is the ratio of the number of males to females. A related concept is what demographers call the marriage squeeze. On average, husbands tend to be a few years older than their wives, so the ratio of men in one age group to women in the next younger age group is an indication of whether or not there is a balance in the number of potential spouses for either sex. The following exercises illustrate these concepts.

1. Choose any nation and compute the 1998 sex ratio for each age group. How does the sex ratio change with age? At what age does the sex ratio become close to 1? Repeat the computation for a second country. Are the trends similar?

2. Propose a hypothesis to explain differences in sex ratio with age. Test your hypothesis by altering the vital rates of a country and looking at the sex ratio of different age groups after 100 years of population change.

3. Using the Tabular Data view for the USA for 1998, compute the ratio of men 20–24 to women 15–29 to women 20–24 and so on, until you reach the last age group. At what ages is the marriage squeeze most pronounced? Compute the USA 1998 sex ratios for each group. Would the marriage squeeze be less of a problem if, on average, spouses were of the same age?

4. Choose and country and modify the fertility rates so that population growth is a large positive value, a large negative value, and ZPG. For each case, simulate 100 years of population growth and examine the sex ratio and marriage squeeze at the end of the 100 years. Is there an effect of population growth rate on sex ratios? Is there an effect of population growth rate on the marriage squeeze? Can you explain these results?

Exponential Population Growth and Decline

If vital rates remain constant, what will happen to population numbers over time? This question is investigated in the following exercises. Keep in mind that although we are working with a human population model, the same principles will apply for any population of plants or animals.

1. Set the number of years to 300 and simulate the population growth of each of the seven nations. Examine the Time Series graphs on both the linear and

logarithmic scales. What happens to each population over the long-term? These simulations assume that current fertility and mortality rates remain unchanged. Is this possible? Why or why not?

2. Exponential growth is described by the following exponential equation:

$$\underline{N}(\underline{t}) = \underline{N}(0)\exp^{rt} \qquad\qquad \text{equation (1)}$$

where $\underline{N}(0)$ is the total population size at some arbitrary initial time, $\underline{N}(t)$ is the population size $\underline{t}$ years in the future, and $\underline{r}$ is the intrinsic growth rate of the population. What will happen if $\underline{r}$ is a positive value? What will happen if $\underline{r}$ is negative? How do the Time Series plots compare with the Growth Rate parameter listed in the summary Stats? Do the populations seem to follow this equation over the long term? If we take the logarithm of both sides of the exponential growth equation, we get:

$$\log \underline{N}(\underline{t}) = \log \underline{N}(0) + \underline{rt} \qquad \text{equation (2)}$$

3. What would you expect from a plot of the logarithm of population size over time? How the case for $\underline{r}{>}0$ differ from $r{<}0$? Examine the Time Series plots of population size versus time for each of the seven nations with the Logarithmic Scale option selected. Do the logarithmic plots of the long-term population values agree with your predictions?

4. Let's use equation (1) to predict population size. Choose any nation and simulate population growth for 300 years. Using the Intrinsic Growth Rate parameter, try predicting the 2298 population size using the population sizes for 1998, 2098, and 2198. (Set $\underline{r}$ equal to the intrinsic growth rate, set $\underline{N}(0)$ to the population size for 1998, 2098, or 2198, and use $\underline{t} = 100, 200,$ or 300 years.) Since the intrinsic rate is displayed to three significant figures, your predictions will be limited to three significant figures of precision. How well does the exponential growth model do with the three predictions? Which predictions are most accurate? Can you explain why this might be the case? Try another nation to see if you get similar results.

5. Another measure of population growth is the doubling time, if $\underline{r}{>}0$ or half-life, if $\underline{r}{<}0$. The doubling time is the number of years it takes the population to double in numbers. The half-life is the amount of time it takes the population to decrease by 50%. The doubling time or half-life is given by the equation:

$$\underline{T}{=}0.6931/|\underline{r}| \qquad\qquad \text{equation (3)}$$

where $|\underline{r}|$ refers to the absolute value of the intrinsic growth rate. If $r{>}0$, T is a doubling time; if $r{<}0$, T is a half-life. Compute the doubling times or half-lives for each of the seven nations. What are the political and social implications of these values?

6. Choose any country, alter its 1998 population size, and simulate the population for 300 years. What effect, if any, does this have on the long-term population growth rates? Repeat the procedure for alterations in the population structure, fertility rates, and female and male mortality rates. Come up with a hypothesis for how changes in each of these factors affect the long-term population growth rates. Design and carry out simulations to test your hypotheses.

7. Try deriving equation (3) from equation (1). [*Hint*: Set $\underline{N}(t)$ equal to $2\underline{N}(\underline{0})$ or $\underline{N}(\underline{0})/2$.]

<u>Group Exercise</u>

Given your observations on the differences in the demographic variables considered in the previous assignments, work in groups of four or five students to develop a hypothesis to consider which nations will exhibit positive long-term growth rates and which ones will have negative long-term growth rates. For each nation, click the Run button to simulate 100 years of population change and look at the Stats view of the output. The last three statistics are all measures of long-term population growth rates. How well do your rankings agree with the rankings of the actual measures of growth rate? Compare and contrast the summary statistics for the different nations.

Pick one of the nations that exhibits a positive long-term growth rate and discuss the following:

1. Should this nation take measures to limit its population size? If so what should those measures be? For example, should a nation limit immigration?

2. Should this nation use either voluntary or mandatory contraceptive programs to limit population size?

3. Identify environmental, health, and other quality-of-life issues that may arise as a result of long-term population growth.

4. Prepare a plan that could be used reduce population growth. If you were the leader of this country, how would you justify your plan to the citizens of your country?

Once you have discussed these topics within your group, compare your responses to those of other groups of students who evaluated the same country.

EvolutionLab

Background

Evolution is the study of how modern organisms have descended from the earliest life-forms and of the genetic, structural, and functional modifications of a population that occur from generation to generation. The ability of a population of organisms to respond to change in their environment and survive and reproduce by developing the characteristics or modifications necessary for survival is known as **adaptation**. Understanding how life evolves is a central concept in biology. The incredible diversity of living organisms that exists and their adaptations to their environment are the direct result of evolution that has occurred over very long periods of time.

Many scientists have contributed to our modern-day understanding of evolution. One of the early pioneers of evolution was the French biologist Jean Baptiste de Lamarck. Working in the early 1800s, Lamarck compared fossilized forms of invertebrate organisms and noted that younger fossils showed advanced structural characteristics compared with older fossils. Lamarck suggested that younger fossils displayed adaptations, or modifications of structures, that were indicative of an increase in an organism's complexity. He proposed that these modifications arose from the use or disuse of body parts. He also hypothesized that certain structural features developed in response to an organism's environment and that many of these acquired characteristics could be inherited by offspring.

One classic example of such modifications suggested by Lamarck involves the long neck of giraffes. He proposed that the earliest giraffes were relatively short-necked animals that were forced to stretch their necks to eat their desired food, leaves that grew at the tops of trees. Over time, such stretching would produce animals with small increases in the length of their necks. This trait would be passed on to future offspring, which would also show increased neck length as they searched for food. Over many generations, these adaptive changes would produce giraffes with a greatly increased neck length compared with their early ancestors. At the time, however, a biological basis for Lamarck's hypotheses remained unclear.

The English naturalist Charles Darwin is generally recognized as the father of evolution. Many of Darwin's now-famous hypotheses were developed as a result of his 1831 voyage on HMS *Beagle*, a ship commissioned by the British government. In this historic voyage, the *Beagle* traveled to many coastal areas of South America. The location that resulted in Darwin's most important observations was the Galápagos Islands in the South Pacific Ocean, off the northwestern coast of South America. Darwin studied populations of Galapagos tortoises and observed their feeding habits regarding the fruit of prickly pear cactuses. Darwin noticed differences in tortoise variety and prickly pear cactus growth that appeared on the various islands of the Galápagos. For example, he observed that on islands without tortoises, prickly pear cactuses grew fruit very close to and, in some cases, on top of the ground. But on islands with tortoises, the prickly pear cactuses showed tall trunks that elevated the fruits of the cactus out of the reach of ground-dwelling tortoises. These observations led Darwin to consider interrelationships between tortoises and their food source. He also observed that many of the different islands of the Galapagos contained distinctly different varieties of finches. It was from Darwin's observations of finches that many of his pioneering theories of evolution by natural selection were developed. Most of these theories continue to provide the basis for our current-day understanding and study of evolution.

Several years after Darwin's initial voyage to the Galápagos Islands, he developed several hypotheses that he felt explained how diverse species of tortoises and finches may have developed on the different islands. He realized that although the populations of these organisms could grow rapidly, tortoise and finch populations did not increase in an uncontrolled manner; rather, the populations of these animals were somehow maintained in a relatively constant size. Darwin noted that certain individual organisms in a population showed subtle structural and functional differences. Because the habitat on each island appeared to stay the same without dramatic changes in environmental conditions such as food supply, he reasoned that there would be competition among individuals for these conditions. In response to this competition, Darwin suggested, some individuals in a population were likely to develop certain structural and functional characteristics, or traits, as a way to increase their odds of survival. Developing the traits that allow individuals to survive and reproduce in an environment is called **adaptation**. Assuming that certain structural differences were essential for survival, he reasoned that these organisms would be superior at surviving in an environment where a population was competing for limited resources. Such favorable structural differences would be maintained or inherited over many generations, while less favorable structural differences were likely to be lost. This line of thinking is often referred to as survival of the fittest because, over time, a population would consist of organisms that inherited those favorable traits that are best suited for survival of the population.

Darwin proposed that evolution of a population in this fashion occurred by **natural selection**, because the environment selects for individuals with traits that allow them to adapt to a given environment.   Because of the greater **fitness** of these organisms, they would be more likely to produce the greatest number of offspring, whereas those organisms that possessed less favorable abilities would be more likely to die. The term fitness is used to describe the reproductive success of an organism. Environmental factors that influence adaptation can be nonliving (**abiotic**) environmental components or living (**biotic**) components, such as competition of organisms for a food source or mating behaviors of a population. Over many generations, natural selection results in changes in the individuals of a population that allow the population to evolve. It is important to realize that natural selection happens to individuals, but that evolution is a function of the changes of the collective individuals in a population over time and not a set of changes that are observed in a given individual.

Darwin's pioneering observations and hypotheses were chronicled in his famous book, *On the Origin of Species by Means of Natural Selection,* which was originally published in 1859. Around the same time that Darwin was developing his important theories, Alfred Russel Wallace was also independently formulating lines of reason that supported evolution by natural selection. At the time of Wallace's and Darwin's work, and for many years afterward, the significance of evolution by natural selection was not fully realized by biologists. In more recent times, however, the concept of evolution by natural selection has been substantiated and validated by the evidence provided from fossil records, ancestry studies of animals that consider embryological stages of development in related organisms, comparative anatomy, and most recently, DNA sequence comparison of related organisms.

The biological basis for the inheritance of traits and the principles of evolution established by Lamarck, Darwin, Wallace, and others was significantly advanced by the work of the Augustinian monk Gregor Mendel. Mendel studied inheritance of traits using the garden pea, *Pisum sativum*. In the 1860s, remarkably, without an understanding of chromosomes and genes, Mendel established that units of inheritance existed to transmit information on traits from parents to offspring. Many years later, the **gene** was identified as the unit of inheritance. Mendel also established the fundamental rules and patterns by which traits are inherited that continue to form the basic principles of genetics that are followed in modern-day genetics laboratories. Mendel's discoveries, and subsequent work on gene structure and function, established an explanation for how organisms can change over time to produce individuals with desirable heritable features that increase the odds for population survival. Mendel recognized that organisms typically contain alternate forms of genes, called **alleles**. Within a given population, several different alleles for a given trait may exist (for example, eye color as a trait is determined by combinations of different alleles for the protein melanin). One way that alleles arise is through mutations—changes in the nucleotide sequence of a gene. Mutations are largely responsible for the production of new alleles and the tremendous genetic diversity that exists in virtually all populations of organisms. The sum total of all genes in a population is defined as a population's **gene pool**. The spread of new alleles within a population and between different populations forms the molecular basis for evolution by natural selection.

Changes in the **genotype** of an organism as mutations arise create changes in **phenotype**—the behaviors, structural appearances, and physiology of an organism. Thus, natural selection produces a range of different genotypes and phenotypes in a population over time. Population genetics involves studying the frequency and distribution of gene inheritance within a population and the genetic variations that arise within populations. Evolution by natural selection results from generational changes in population frequencies of alleles that influence the phenotypic characteristics of a population. Allele frequency is a measure of the abundance of different alleles in a population. Evolution occurs over many generations as individuals with different alleles reproduce. As individuals in a population breed with each other and with individuals from other populations, allele frequency will change as each population gains some alleles and loses others. This concept is known as **gene flow**. Hence, evolution results from inevitable changes in large numbers of genes within a population over many generations. There are three modes by which the heritability of a trait in a population can be influenced by natural selection: **directional selection**, **stabilizing selection**, and **diversifying selection**. These modes are discussed in the Assignments sections of EvolutionLab.

Evolution is a central theme that unifies and connects virtually all disciplines of biology. However, because evolution occurs over long spans of time and is difficult to observe directly, the mechanisms that drive natural selection do not lend themselves to experimentation during a few hours or even a semester spent in the biology lab. In EvolutionLab, you will have the opportunity to simulate experiments in evolution to help you understand how traits of an organism can change in response to different biological and environmental conditions. You will study how beak size and population numbers for two hypothetical populations of finches on two different islands can evolve in response to factors that you will manipulate by changing environmental conditions on these islands. EvolutionLab is not an exact simulation of finch beak evolution; rather it is a model designed to demonstrate important principles of adaptation by natural selection. Later you will have the opportunity to study Mendelian genetics in the fruit fly, *Drosophila melanogaster*, using FlyLab. You will also have the opportunity to use PedigreeLab to study how genetic traits are inherited across several generations of humans.

References
1. Darwin, C. *On the Origin of Species by Means of Natural Selection, or the Preservation of Favored Races in the Struggle for Life.* New York: New American Library, 1963.

2. Gould, S. J. "The Evolution of Life on the Earth." *Scientific American*, October 1994.

3. Grant, P. R. "Natural Selection and Darwin's Finches." *Scientific American*, January 1991.

4. Weiner, J. *The Beak of the Finch: A Story of Evolution in Our Time.* New York: A. Knopf, 1994.

Introduction
EvolutionLab will allow you to study important principles of evolution by examining small populations of finches on two different islands, "Darwin Island" and "Wallace Island." You will manipulate important parameters that influence natural selection and then follow how your changes influence the evolution of beak size and population numbers for the two different populations of finches over selected time intervals. One obvious advantage of this simulation is that you can quickly and easily create new parameters and model evolutionary changes over long time periods.

Objectives
The purpose of this laboratory is to:
- Help you develop an understanding of important factors that affect evolution of a species.
- Demonstrate important biological and environmental selection factors that influence evolution by natural selection.

- Simulate how changes in beak size and other characteristics of finch populations influence evolution of beak size and population numbers.

Before You Begin: Prerequisites

Before beginning EvolutionLab you should be familiar with the following concepts:

- Evolution and the importance of natural selection as a central concept of Darwinism (see Campbell, N. A., Reece, J. B., and Mitchell, L. G., *Biology* 5/e, chapter 22).
- Biological and environmental factors that influence natural selection (chapter 23).
- Evolutionary principles associated with population genetics (chapter 23).

Assignments

The finches on Darwin and Wallace Islands feed on seeds produced by plants growing on these islands. There are three categories of seeds: soft seeds, produced by plants that do well under wet conditions; seeds that are intermediate in hardness, produced by plants that do best under moderate precipitation; and hard seeds, produced by plants that dominate in drought conditions. EvolutionLab is based on a model for the evolution of quantitative traits—characteristics of an individual that are controlled by large numbers of genes. These traits are studied by looking at the statistical distribution of the trait in populations and investigating how the distribution changes from one generation to the next. For the finches in EvolutionLab, the depth of the beak is the quantitative trait. You will investigate how this trait changes under different biological and environmental conditions.

You can manipulate various biological parameters (initial beak size, heritability of beak size, variation in beak size, clutch size, and population size) and two environmental parameters (precipitation, and island size) of the system, then observe changes in the distributions of beak size and population numbers over time.

Getting to Know EvolutionLab: The Influence of Precipitation on Beak Size and Population Numbers

1. The first screen that will appear in EvolutionLab presents an initial summary (Input Summary) of the default values for each of the parameters that you can manipulate. Notice that default values on both islands are the same. Click on the Change Inputs button at the left of screen to begin an experiment. A view of initial beak size will now appear. In the Change Input view you can change the biological and environmental parameters in EvolutionLab to design an experiment. This first experiment is designed to study the influence of beak size on finch population numbers. For finches, deep beaks are strong beaks, ideally suited for cracking hard seeds, and shallow beaks are better suited for cracking soft seeds. Develop a hypothesis to predict how changes in beak size will affect population numbers for these finches. Test your hypothesis as follows:

   a. Begin by setting the initial beak size on the two islands to opposite extremes. Leave the initial beak size on Darwin Island at 12 mm and click and drag on the slider to change the initial beak size on Wallace Island to

28 mm. Note the change in beak size that appears on the graphic of each finch. Click the Done button to return to the input summary view. Notice the new input value (28 mm) for beak size on Wallace Island while beak size on Darwin Island remains at the default value of 12 mm. Use the popup menu in the lower left corner to select a value of 300 years, and run the simulation by clicking the Run Experiment button.

    b.   Once the experiment has run, you will be in the Beak Size view. Look at the plots of average beak size over time. What do you observe? Do you notice any trends in beak size? Click on the Population button and look at the plots of population numbers over time. What changes do you see? Do the two islands differ? Does the data support or refute your hypothesis? Data from the Beak Size view, and Population view are shown in tabular form in the Field Notes view. Click on the Field Notes view. A table showing each year of the experiment, mean beak size, and population will appear.

        Click on the Histograms button. These are plots of surviving birds and total birds plotted against beak size. Click and drag the slider to advance the years of the plot and to see how beak size on each island may have changed over time. Note how the distributions of beak size change over time. What happened to beak size on Darwin Island compared to Wallace Island over time? Is this what you expected? Why or why not? Click on the Input Summary button to see a table of your input values for this experiment. Data from each of the views that you just looked at can be exported by clicking on the Export button. A separate window will now open. You can print your data directly from this window or save this data to a disk.

2.   This experiment is designed to explore the effect of precipitation on finch beak size and population numbers. Click the New Experiment button, click the Change Inputs buttons then click the Precipitation button. Recall the relationship between precipitation and seed growth. There are three categories of seeds: soft seeds, produced by plants that do well under wet conditions; seeds that are intermediate in hardness, produced by plants that do best under moderate precipitation; and hard seeds, produced by plants that dominate in drought conditions. Develop a hypothesis to consider how a decrease in precipitation on Darwin Island might affect beak size and develop a hypothesis to explain how a decrease in precipitation might influence population numbers for these finches over time. Test your hypotheses as follows:

    a.   Leaving all other parameters at their default values, decrease precipitation on Darwin Island to 0 mm. Notice how the distribution of seeds produced on Darwin Island changes as you change precipitation. Set the experiment to run for 100 years, then run the experiment. Compare beak size and population numbers for the finches on Darwin Island over 100 years. Scroll down the Field Notes view to observe the data recorded over 100 years. Use the Beak Size and Population buttons to view the effect of your experiment on each of these parameters. Did you notice any trends in the distributions of beak size? What did you observe? Did you notice any trends in population number? What did you observe? Explain your answers. Run another experiment for 200 years by clicking on the Revise Experiment button. Use the popup menu at the lower left corner of the screen to select a value of 200 years, then click the Run Experiment button. Repeat this experiment for 300 years. What changes did you observe in beak size and population numbers? Do these results confirm or refute your hypothesis? If necessary, reformulate your hypothesis and test this hypothesis.

b. Perform the same experiment for both Wallace Island and Darwin Island simultaneously. Did you notice any differences between precipitation, changes in beak size, and population numbers for the finches on Wallace Island compared with those on Darwin Island? Explain your answers.

c. Develop a hypothesis to consider how an increase in precipitation on Darwin Island might influence the evolution of beak size. Click the New Experiment button, return to the Change Inputs view then increase the precipitation on Darwin Island fourfold while leaving precipitation on Wallace Island at the default value. Run this experiment for 300 years to test your hypothesis. What did you observe? After you have observed the data for this experiment, rerun this experiment. Look at the output results in the Beak Size and Population views. Do you notice any differences in this rerun compared with the previous run? Are the general trends observed in this run the same as the previous run? Explain your answers. Run and rerun each experiment for 100, 200, and 300 years. Perform another experiment to test your hypothesis by increasing precipitation on Wallace Island to 50 cm/year and increasing beak size to 28 mm. Run an experiment for 300 years and describe your results. Do these results support your hypothesis?

d. Leaving beak size at the default value, equally decrease precipitation to a value close to 0 cm on both islands. Run an experiment for 200 years and carefully compare the effects of decreased precipitation on the distribution of beak size and population numbers for the finches on each island. Rerun this experiment several times. Provide explanations for any differences in beak size and population numbers that you observed when comparing the finches on both islands.

Modes of Natural Selection

There are three primary ways by which natural selection can influence the variation of a trait in a population. The three modes of selection are **directional**, **stabilizing**, and **diversifying**. In directional selection, changing environmental conditions can favor individuals with phenotypes that are at opposite extremes of the distribution range for a given trait (for example, finches with very deep beaks or finches with very shallow beaks). Stabilizing selection, unlike directional selection, selects against individuals with extreme phenotypes and favors individuals with more intermediate or average values for a given trait. Disruptive selection favors individuals at both extremes of the distribution range for a trait while selecting against individuals with average values for the trait. The following assignment is designed to help you understand how environmental changes can result in different modes of natural selection.

1. Leaving all of the other settings at their default values, change the rainfall on Wallace Island to 50 cm/year and the rainfall on Darwin Island to the minimum possible value (0 cm/year). Run the simulation for 300 years.

   a. Look at the plot of beak size over time. What type of selection is taking place on Wallace Island? On Darwin Island? Explain your answer.

   b. Using the numbers from the field notes, calculate the average R (the difference in mean beak size from one generation to the next) in the first ten years of the simulation and the last ten years for both populations.

   c. Look at the plot of finch population over time. Explain the reason for any differences in population numbers between the two islands.

Effect of Clutch Size

Clutch size is the number of eggs that a female bird lays in her nest. In the EvolutionLab simulation model, birds mate for life and live for one year, and each female produces only one clutch of eggs per year. For a mating pair to replace themselves, they must produce at least two offspring; this is why clutch size is set to a minimum of two eggs. The maximum clutch size of thirty eggs is unrealistic; however, most bird species produce more than one clutch in their lifetime, so a total of thirty or more offspring per mating pair is possible. You can use the sliders to change the mean clutch size for each island population. As you move the slider, the number of eggs in the nest will change to reflect the value that you have chosen.

1. Set the clutch size to 6, the rainfall to 37 cm, and the initial beak size to 25 mm on both islands (keep everything else at the default values). Run this simulation for 300 years and repeat the run two or three times. Did you notice anything odd? (If not, try again until you do). Propose a hypothesis to explain this result. Leaving the rainfall and beak sizes alone, what parameters would you change to prevent this? What parameter would you change to increase the likelihood of this happening? What type of selection is happening during the first several years of this experiment? If these birds were capable of **assortative mating**, what might happen on one island?

Effect of Island Size

The size of the living area for any population can strongly influence population numbers for organisms that live within that environment. The maximum number of organisms from a given population that an environment can support is known as the **carrying capacity** of that environment. Island size is one factor that can determine the carrying capacity of finches on each island. For the purposes of this simulation, the islands are assumed to be roughly circular and island size is represented as the radius of the island in kilometers. The size of each island remains constant throughout the simulation unless you choose to change this parameter. Although changing the entire size of an island is not something that could easily be done in real life, habitat changes and reducing the living environment for a population are real changes that occur through processes such as land development, and pollution. The following assignment is designed to help you learn about the influence of island size on the carrying capacity of finches on Darwin and Wallace Islands.

1. Develop a testable hypothesis to predict what effect an increase in island size will have on beak size and finch populations. Test your hypothesis as follows, Begin your experiment by leaving all other parameters at their default values. Select the Island Size input and use the sliders to increase the size of either Darwin or Wallace Island. As you move the slider, the island image will change to reflect the values you have chosen. What effect did this change in island size have on finch population? What effect did this change in island size have on beak size? Are the results what you expected? Explain your answers. Perform a new experiment to learn about the effects of a decrease in island size on beak size and finch populations.

2. Based on the previous experiments, consider possible parameters that you could manipulate which would prevent the changes in population size and beak size that you observed from occurring. Test the effect of these parameters to influence population size and beak size by designing and running experiments to confirm or refute your answers.

Variance

Variance is a measure of how different the phenotype is from one bird to the next. If the variance of a trait is large, then there will be large differences in phenotypes among birds for a given trait. If, on the other hand, variance is low (near zero), then all of the birds will be very similar to one another. This parameter determines two values for the simulation. First, it determines the initial *population variance—*variance in beak size for the entire population. The population variance may change during the course of the simulation. The variance parameter also determines the *sibling variance—*variance in beak size among individuals with the same parents. The sibling variance is a measure of how much variability is inherent in a trait. If you look at the plots of offspring beak size versus the midparent value for the heritability parameter, the sibling variance represents the amount of dispersion of the points around the regression line. In EvolutionLab, the sibling variance remains constant throughout the simulation. You can use the sliders to change the variance for each island population. As you move the slider, the width of the bell-shaped probability distribution will change to reflect the value that you have chosen.

1. For the finches on one of the islands, develop a hypothesis to consider the effects of changes in variance on population numbers. Begin by decreasing variance to a value close to zero. Run your experiment for 300 years. Explain the results of this experiment. Perform a similar experiment for an increase in

variance.   Describe your results.   Can you reverse any of the trends in
population number by changing other variables such as precipitation?

Extinction

Now that you have manipulated many of the biological and environmental factors in
EvolutionLab, consider how these factors could lead to extinction of the finch
population on either island.

1. What conditions would lead to extinction of a finch population? Which of the
    parameters is most important in determining whether a population goes
    extinct? Can you describe at least three different sets of conditions that will
    lead to extinction?

2. Design and perform experiments that will confirm or refute your answers.
    Compare your answers with those of other classmates. How do they compare?
    If your classmates have different conditions that lead to extinction, design and
    perform an experiment to validate the ability of those conditions to cause
    extinction.

Group Assignment: Influence of Heritability

Heritability is a measure of the genetic contribution to a phenotype of an organism. If
the heritability of a trait is large (near one), then the progeny of a mating will be more
similar to their parents than unrelated individuals. If heritability is low (near zero),
then the environment will be more important in determining the phenotype and the
progeny will vary around the average value for the population. One method for
measuring heritability is to plot the beak depth of each bird versus the average beak
depth of each bird's parents (midparent value). The slope of the line produced by this
plot represents heritability.

You can use the sliders to vary the heritability of beak size on each island population.
Heritability remains constant at this value throughout the simulation. As you move
the slider, the graph of offspring beak size versus the midparent value will change to
illustrate the heritability value that you are selecting. The mean value for the plot is
equal to the initial mean beak size that you have chosen. The amount of dispersion of
the points around the plot is determined by your selection for variance in beak size.

Work together in a group of four students to perform the following assignment, which
is designed to help you understand the influence of heritability on population numbers.
Click on the Heritability button and read the description of heritability and heritability
plots.

1. What effect do you think heritability will have on the evolution of finch beak size?
    As a group, discuss the importance of heritability. Design at least one hypothesis
    to explain how changes in heritability might affect the evolution of beak size and
    then use EvolutionLab to test your hypotheses. What did you do, and what were
    the results?

a. Looking back over the experiments you did with heritability and variability, would you say that changes in these properties of the populations cause similar effects or different effects on evolution of the finches' beaks? Does either of these parameters affect the direction of evolution? The end point of evolution? The rate of evolution? Discuss your answers with students in your group.

b. If the heritability of one population is set to 0.75 while the other is set to 0.25, could the rate of evolution on the two islands still be the same if there were differences in variability between the two islands? Try it and see if you can find values for variability that compensate for the differences in heritability. What is going on here and what is the connection between heritability and variability?